m
MEDIA MANUALS

Basic TV Technology:
Digital and Analog
Third Edition

m
MEDIA MANUALS

Basic TV Technology: Digital and Analog

Third Edition

Robert L. Hartwig

Focal Press

Boston Oxford Auckland Johannesburg Melbourne New Delhi

Focal Press is an imprint of Butterworth–Heinemann.

Copyright © 2000 Butterworth–Heinemann

 A member of the Reed Elsevier group

Recognizing the importance of preserving what has been written,
Butterworth–Heinemann prints its books on acid-free paper whenever possible.

 Butterworth–Heinemann supports the efforts of American Forests and
the Global ReLeaf program in its campaign for the betterment of trees,
forests, and our environment.

Library of Congress Cataloging-in-Publication Data
Hartwig, Robert L.
     Basic TV technology: digital and analog / Robert L. Hartwig.—3rd ed.
          p.   cm.—(Media manuals)
     Includes bibliographical references.
     ISBN 0-240-80417-1 (pbk.: alk. Paper)
     1. Television—Handbooks, manuals, etc. I. Title. II. Series.
TK6642.H37   2000
621.388'8—dc21   99-054231

British Library Cataloguing-in-Publication Data
A catalogue record for this book is available from the British Library.

The publisher offers special discounts on bulk orders of this book.
For information, please contact:
Manager of Special Sales
Butterworth-Heinemann
225 Wildwood Avenue
Woburn, MA 01801-2041
Tel: 781-904-2500
Fax: 781-904-2620
For information on all Butterworth-Heinemann publications
available, contact our World Wide Web home page at:
http://www.bh.com

10 9 8 7 6 5 4 3

Printed in the United States of America

Contents

viii

Introduction

There was some debate between author and editors as to whether this edition should eliminate all reference to analog technology. It is clear that the video industry has made large strides toward a full digital environment. Many medium and small market TV stations, cable operations, schools, and businesses still have a good deal of analog equipment. It is for this reason we decided to make this a book that covers digital technology and analog technology. It is intended to be a transition piece for this transition period.

This is not a TV production textbook, but it is for TV production people. This book doesn't deal with TV production techniques or TV audio. Several fine books already cover those subjects. I see no reason to write what others have already done a better job of writing. Rather, this book deals with two interrelated subjects. I hope to show you how the various pieces of video equipment are integrated to form a complex video system. But to understand that, you must first have some knowledge of how the equipment works and what goes on inside it.

As TV equipment becomes more complex and sophisticated, it becomes more important to understand how that equipment works. This is especially true in the worlds of instructional and industrial TV, where one person may have to do it all. Having an understanding of how the equipment and systems work gives you two distinct advantages. First, you will be more adaptable to different makes, models, and features. New buttons and knobs aren't as likely to intimidate you. Second, you can be more creative. You're not limited to what you have been shown, but can figure out new applications and how to solve problems for yourself.

One need not be an engineer or know advanced math and physics to understand the basics of how TV equipment works. After teaching much of this material for more than 20 years, I know that students with little or no math and science background can be taught to understand the equipment. However, students must realize that, because of their lack of background, they may not be able to get detailed answers to all of their questions.

Based on feedback from former students and their employers, I'm confident that the material in this book is the single most important body of material that my TV students receive. Many of them feel that it's even more important than experience using the equipment.

Since this book deals with television systems, it's difficult to understand some topics without the proper foundation. This book, then, uses the building block approach. Many topics rely on information from previous topics. Most of this book should be studied sequentially.

I've attempted to make this book as easy to read and understand as possible, but it should be recognized that few of us can learn TV production or TV systems just from a book. The best learning will take place if you use this book in conjunction with hands-on TV production experience.

Everything from directing to shading to tape operation will help make the contents of this book more meaningful. I've tried to make the text clear, concise, and conversational in nature.

I have, on occasion, simplified the facts and various theories somewhat in order to make some concepts a little clearer. I hope more knowledge-able readers will not oppose these changes.

Acknowledgments

The contents of this book have been accumulated over a period of years. Much of this technical information was generously supplied by patient engineers who were willing to take the time to share their knowledge of their fields with one who was less knowledgeable.

Among them were the engineering staff at the TV studio of California State University, Chico, who, in my student days, were always willing to answer the technical questions that weren't covered in class. The man who designed and supervised the building of our studio at Cuesta College, Darrell Wenhardt of Creative Broadcast Techniques in San Diego, always found the time to discuss and explain emerging technologies, even years after his contractual obligations to the school had ended. Finally, I've been fortunate with regard to the maintenance engineers I've known at Cuesta College. Ken James, now with the Grass Valley Group, and Jan Schaafsma, now a development engineer at Harris Corporation, and currently Bill Bordeaux spent many hours of discussion helping me to expand my knowledge of this field.

Two people, in particular, also read this manuscript and made suggestions, both major and minor, that have contributed to this project. Professor Donald R. Mott of Butler University carefully read the first edition manuscript and his detailed comments contributed greatly and made that edition a better book. I hope that this edition follows in the footsteps he helped direct. Darrell Wenhardt of Creative Broadcast Techniques has come through with many appreciated comments and suggestions for every edition of this book.

A special thanks goes to the people at Focal Press. When I started the first edition of this book, I had no idea how much editors, production people, and others added to the finished product; I thought it was the author who made the book. I know better now. The questioning, prodding, suggestions, refining, and editing by the people at Focal have made this a much better work than the one I originally created.

The Atom and Electricity

To understand the technical aspects of television systems and equipment, it is important to know some basic theories, terms, and abbreviations in the field of electronics.

The parts of the atom

The first thing to do is get a basic understanding of how electricity works. If you think back to your high school science classes, you'll recall that an atom is made up of three parts. The neutron is in the center or nucleus of the atom and has no electrical charge. Protons, also in the nucleus of the atom, have a positive charge (+). Like the planets circling the sun in the solar system, electrons circle the nucleus, and have a negative charge (−). Since there are the same number of protons and electrons, the atom as a whole has no electrical charge.

The flow of electrons through metals

In some elements, usually the elements that we call metals, electrons can be very easily dislodged from their orbits. When they are knocked out of their orbits, they are attracted to other atoms and knock these atoms' electrons from their orbits. This flow of electrons is electricity. Of course, the atoms that have lost electrons now have an overall positive charge and tend to attract the loose electrons.

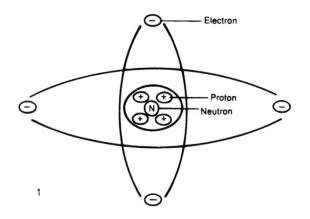

Electron

Proton

Neutron

N

1

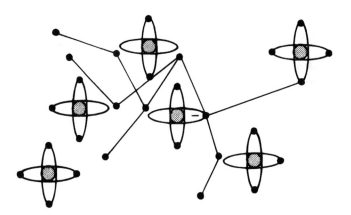

2

1. The parts of the atom.

2. The flow of electrons.

All complete circuits are loops of flowing electrons.

Basic Circuits

Direct current (DC)

A flashlight battery has the ability to provide a flow of electrons, but accomplishes nothing sitting on the shelf, since it is only stored energy. Only when the positive terminal of the battery is attached to one end of a light bulb and the other end of the light bulb is attached to the negative terminal of the battery do we have a completed circuit, and the light bulb emits light. In this circuit all of the electrons continue to flow in the same direction. This is called *direct current* (DC). Of course, if there's only one wire between the light bulb and the battery, or if there's an open switch, the circuit is incomplete. The electrons have to flow through a complete loop in order to do work.

Such is the case with all electrical circuits. There must be a complete loop providing both a place for electrons to come from and a place for them to go. This is why the plugs to all your appliances at home have two prongs. Most electrical devices will also have a switch somewhere in the circuit to allow you to interrupt the flow of electrons.

Alternating current (AC)

The preceding example explained a simple DC circuit. Most circuits used in video equipment use direct current. However, there is another type of current that you will commonly encounter called *alternating current* (AC). With AC the flow of electrons changes direction constantly. The current flows from negative to positive, then from positive to negative, and so on. AC is much more efficient for transmission through wires over long distances. That's one of the reasons that it is used for household electricity. Some household appliances work better on AC (your clothes washer, for example) and others work better on DC (like the internal circuits of your TV), so the ability to easily change AC into DC is another good reason why we use AC power for our main electricity supply.

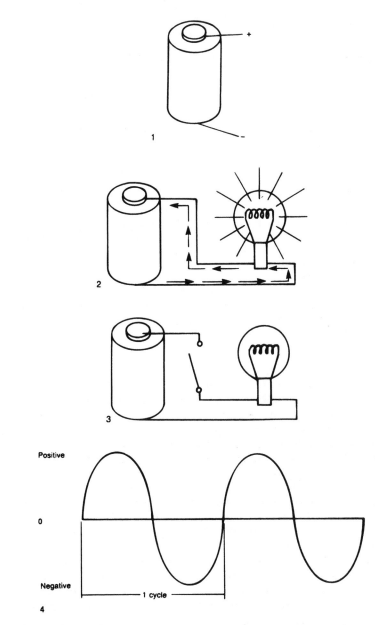

1. Battery (stored energy).

2. Simple circuit.

3. Simple circuit with switch.

4. Alternating current.

When you're dealing with electricity, you need to be able to measure it.

Units of Measurement (1)

Voltage

There are several types of measurements relating to electricity. The first of these is *voltage*, which is measured in *volts* (V). Voltage is the pressure of the electricity. In a given medium, the speed of electricity is constant, but the pressure is not. The typical flashlight battery has a voltage of 1.5 V. The electricity in your home, on the other hand, is 120 V. So you could say that the electricity in your house has a lot more pressure behind it.

Current

The second area of measurement for electricity is *current*, which is measured in amperes or amps (A). Current is the volume of electrons, that is, the number of electrons passing a certain point in a given time. A current of 4 amps has twice as many electrons passing by as does a current of 2 amps.

Power

Voltage and current together determine *power*. Power is a measurement of work being accomplished and is measured in *watts* (W). Watts are determined by multiplying volts by amps. For example, if you have a light bulb that draws 1.25 amps and the house voltage is 120 V, your light bulb is a 150-W bulb (1.25 amps × 120 V = 150 W).

Resistance and impedance

Another area of measurement is *resistance*. Any DC electrical circuit will resist the flow of electrons. We measure this resistance in *ohms* (Ω). Closely related to resistance is *impedance*. Impedance can help tell the production person if two or more circuits will interact well. The following oversimplified example may help you understand the concept. If your stereo amplifier has a speaker impedance of 8 Ω, this means that it is designed to hook up to speakers that have 8 Ω of resistance. If you connect your 8-Ω amplifier to your 8-Ω speakers, everything works great. But what happens when you connect that 8-Ω amplifier to speakers that have 10,000 Ω of resistance? Not much! The system just isn't designed to overcome that much resistance. On the other hand, if you have both an amplifier and speakers with 10,000 Ω of impedance, everything works just fine. But if you connect that 10,000-Ω amplifier to speakers that have 8-Ω resistance, you've got problems. You could destroy your speakers! They're just not designed to work with that amplifier. You have what's called a *mismatch*. Impedance is an important factor when integrating electrical components.

Mathematical symbols and formulas

Because they all concern the flow of electricity through a conductor, these basic units of measurement are all mathematically related. In addition, when working with units of measurement mathematically, we give them different symbols. The mathematical symbols and the basic formulas are shown in the table. You may not need to memorize these formulas, but you should know that they exist and that the units are interrelated. It is also important to remember the different mathematical symbols.

6

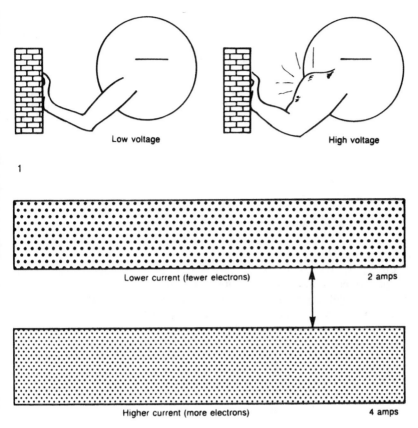

Low voltage

High voltage

1

Lower current (fewer electrons)   2 amps

Higher current (more electrons)   4 amps

2

1. **Voltage.**

2. **Current.**

Mathematical symbols and formulas.

Unit	Mathematical symbol	Basic formulas
Watts	W	$E \times I$ or $I^2 \times R$ or $\dfrac{E^2}{R}$
Volts	E	$I \times R$ or $W \times R$ or $\dfrac{W}{I}$
Amperes	I	$\dfrac{E}{R}$ or $\dfrac{W}{R}$ or $\dfrac{W}{I}$
Ohms	R	$\dfrac{E}{I}$ or $\dfrac{E^2}{W}$ or $\dfrac{W}{I^2}$

There are also other measurements you need to know.

Units of Measurement (2)

Voltage, current, and resistance are basic measurements of electricity, but when you need to apply electricity to television, there are some other measurements you'll need to know. Among these are frequency, hertz, and AC frequency.

Frequency
Frequency is an action that repeats itself. If you have an electrical circuit that puts out repeated and equal bursts or pulses of energy at 100 of those pulses a second, the frequency of that circuit is 100 pulses per second. But we measure frequency in *hertz* (Hz) so the frequency is 100 Hz.

AC frequency
In our earlier discussion of AC you learned that the flow of electrons in AC current constantly changes direction. If the electricity in your home is 120 V AC, what happens is that the electricity goes from 0 V up to +120 V, back down to 0 V, continues down to −120 V, and then goes back up to 0 V. This alternation between 120 V of positive electricity and 120 V of negative electricity is one cycle. Your household electricity does this 60 times a second. So the frequency of your household electricity is 60 Hz. Thus to be fully descriptive of the electricity in your house, you would say it's 120 V 60 Hz AC.

Capacitance
Another term you might hear the engineering staff of the station discussing is *capacitance*. A capacitor is a device that can store an electrical charge, and the storage of that charge is called capacitance. The unit of measurement for capacitors is called the *farad* (F). Capacitors are used for filtering signals and they are used in power supply circuits to help in the conversion of AC to DC.

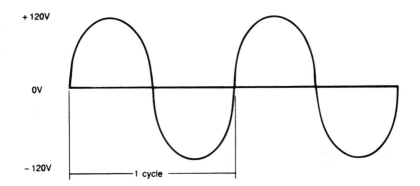

Alternating current.

Abbreviations

Kilo

K is the abbreviation for kilo, which equals 1,000. If something has a frequency of 10 KHz, it has a frequency of 10,000 Hz. You can just replace the word *thousand* with kilo. Likewise, if you have a light in your TV studio that uses 1 KW of power, it is using 1,000 W of power.

Mega

M stands for mega, or 1,000,000. So a generator that puts out 1 MW of power puts out 1,000,000 W of power. If your favorite radio station has an assigned frequency of 96 MHz, it has a frequency of 96,000,000 Hz.

Giga

G stands for giga and is equal to 1,000,000,000. So a measurement of 6 GHz would be equal to 6,000,000,000 Hz.

K, M, and G are used for units that are greater than one. There are, however, several abbreviations that are used for measurements that are smaller than one.

Milli

The first of these is *m*, which stands for *milli* and means 1/1,000. If you took a measurement and it read 5 mV, that is read as five millivolts or five one thousandths of a volt. A measurement of 321 mA would be read as 321 milliamps or 321 one thousandths of an amp.

Micro

The next abbreviation is μ, which stands for *micro* and means 1/1,000,000. Thus, a reading of 25 μsec is read as 25 microseconds or 25 one millionths of a second.

Nano

The last abbreviation is *n*, which stands for nano and means 1/1,000,000,000. So a measurement of 63 nsec would be read as 63 one billionths of a second.

Conversions

You also need to be able to convert these abbreviations. For example, you need to be able to convert KW to MW and vice versa. It's really a simple task, but there are a couple of rules that will make it easier.

To convert 2,575 KW into MW, first ask, "Am I going from smaller units to larger units, or from larger units to smaller units?" Since KW are smaller than MW, you're going from smaller to larger; therefore you move the decimal point to the left. How far to the left? Looking at the top chart at the right you will find the difference in the number of zeros between K and M, and that's how many spaces you move the decimal point to the left. Since M has six zeros and K has three zeros, you move the decimal point three places to the left. As a result, 2,575 KW becomes 2.575 MW. Going from larger to smaller units works just the opposite way. This process is summarized by the charts at the right.

```
G  = giga  =   1,000,000,000
M  = mega  =       1,000,000
K  = kilo  =           1,000
     units =               1
m  = milli =           1/1,000
μ  = micro =       1/1,000,000
n  = nano  = 1/1,000,000,000
```

1

Smaller to larger, decimal point goes left.

Larger to smaller, decimal point goes right.

The difference in the number of zeros determines the number of spaces the decimal point is moved.

2

1. Chart of abbreviations.

2. Rules of conversion.

Induction and Noise

Induction

There are two other basic theories necessary for understanding television equipment. The first of these is *induction*. Any electrical circuit that has a changing flow of electrons will create an electromagnetic field around itself. For example, if you turned a flashlight on and off several times, the flow of electrons would be starting and stopping and a small electro-magnetic field would be created. However, if you left the flashlight on, the flow of electrons would be continuous and unchanging and there would not be an electromagnetic field. Since the flashlight uses very small amounts of electricity, its field would be very small—almost unmea-surable. But a high-tension power line running cross-country has an extremely strong electromagnetic field. When another circuit is placed within this electromagnetic field, a signal from the more powerful circuit is forced into the weaker circuit. The signal may take the form of static, as when you try to play the AM radio in your car near high-power lines, or it may be actual information, as when you sometimes hear very weak background voices on the telephone.

Noise

Another thing that can create problems is *noise*. To see what noise looks like in video, unhook the antenna and/or cable from your TV. Turn your TV on. What you see is noise! If you happen to be near a transmitter and have your TV tuned to its channel, you'll also see some picture. This noise is obviously an undesirable feature. Too much of it and it interferes with the picture or signal. Inherent in every electrical circuit is a certain amount of this noise. If there is too much noise, then there is a problem. Certainly, if you want to watch TV, you don't want to see any noise.

Signal-to-noise ratio

You need to be able to measure the relationship between the strength of the signal and the amount of noise the circuitry creates. This measure-ment is called the *signal-to-noise* ratio. We use the *decibel* (dB) scale to measure this relationship. The dB scale is a logarithmic ratio. The signal-to-noise ratio is doubled for every 3-dB difference between the strength of the signal and the strength of the noise. For example, if the noise in our system is 0 dB and the signal is 3 dB, then the signal is twice as strong as the noise; if the signal is 6 dB, then it's four times as strong as the noise; if the signal is 9 dB, it's eight times as strong; 12 dB, 16 times as strong; 15 dB, 32 times as strong, and so on. In video, we like to have a signal-to-noise ratio of at least 45 dB.

Sometimes a signal-to-noise ratio is written as –45 dB. It really means the same thing. A 45-dB ratio means that the signal is 45 dB stronger than the noise; a –45-dB ratio means that the noise is 45 dB weaker than the signal— same thing. In real numbers a signal-to-noise ratio of 45 dB means that the signal is over 32,000 times stronger than the noise! That's pretty impressive.

12

1. Induction.

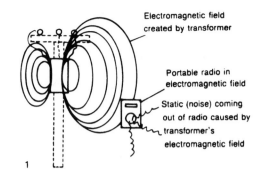

Electromagnetic field created by transformer

Portable radio in electromagnetic field

Static (noise) coming out of radio caused by transformer's electromagnetic field

1

2. Video noise.

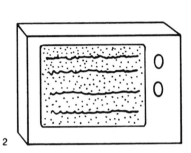

2

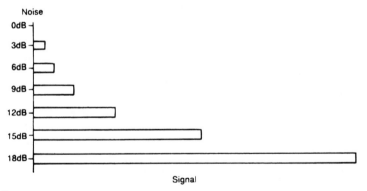

Noise

0dB
3dB
6dB
9dB
12dB
15dB
18dB

Signal

3

+ 45-dB signal-to-noise ratio = signal is 45dB stronger than the noise

– 45-dB signal-to-noise ratio = noise is 45dB weaker than the signal

Therefore, + 45-dB signal-to-noise ratio = – 45-dB signal-to-noise ratio

4

3. Signal-to-noise ratio. Comparison of the strength of the signal with the noise using the dB scale.

4. Minus and positive signal-to-noise ratios.

13

Cathode-Ray Tubes

The television picture tube is properly called a *cathode-ray tube* (CRT). The CRT is a large glass vacuum tube. The inside front of the tube is covered with a phosphorescent substance that glows when struck by a beam of electrons: The stronger the beam of electrons, the brighter the glow; the weaker the electron beam, the less the glow. At the back of the CRT, in the narrow neck, is an *electron gun* (a cathode that is heated) that emits a beam of electrons: The higher the voltage activating the gun, the stronger the electron beam; the weaker the voltage, the weaker the beam. The direction of the electron beam is controlled by the *deflection yoke* (a group of large electromagnets surrounding the middle section of the CRT).

Interlace scanning

Since a beam of electrons isn't very wide, one sweep across the CRT doesn't give us much information. To get more information, the electron beam has to go back and make successive sweeps of the CRT. In the American system, it makes 525 sweeps, or lines, to cover the entire face of the CRT. But things aren't quite that simple. Rather than just sweeping all 525 lines at once, the system goes through and sweeps the odd lines (line numbers 1, 3, 5, 7, 9, . . .) first and then goes back and sweeps the even lines (line numbers 2, 4, 6, 8, . . .). Thus we have two separate *fields* of 262.5 lines each. When these two fields are combined, they give us our single video *frame* or a complete picture of 525 lines. This will give us 60 fields and 30 frames every second. This process is called *interlace scanning* and the reason for it will be explained a little later.

Random interlace

There are two types of interlace scanning. The first is called *random interlace*. In random interlace, the exact position of each line varies with each frame scanned. Certainly, line 5 will be between lines 4 and 6, but it may not be exactly centered between the two lines and its position may vary a little with each frame. This is the system used with home video cameras and, as most of you know from your own experience, it works just fine.

Positive interlace

The other type of interlace scanning is called *positive interlace*. What this means is that each line has a specific location where it will be located in every frame. So line 1 is in the exact same position in every frame, line 2 is in the exact same position, line 3 is in the exact same position, and line 2 will be exactly centered between lines 1 and 3 every time. All professional broadcast equipment uses positive interlace scanning.

You might wonder if you can see the difference between random and positive interlace. If you were to put the two systems next to each other on home TV systems, you probably wouldn't see any difference. The difference starts becoming important when you begin building a system and hooking various components together and trying to use sophisticated production techniques.

14

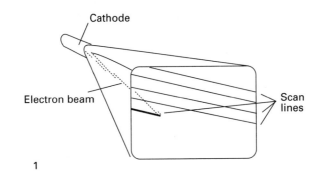

1

2

1. CRT.

2. Analog video system.

The CRT converts the video signal into visual images.

Need for Interlace Scanning

This is a convenient time to explain why the interlace scanning method is used in the system. Our TV system was developed over 50 years ago. Engineers had to work within the limitations of the time. One of those limitations was the phosphor coating of the CRTs. When the CRT's electron beam struck a phosphor on the face plate, it caused that phosphor to glow. As soon as the electron beam left the phosphor, the glow started to get weaker. If our CRT scanned from line 1 all the way down to line 525 in succession, by the time the electron beam got to the bottom of the picture, the top of the picture would be pretty dark. To prevent this, the electron beam scans a field of 262.5 lines and then goes back to the top of the picture. Just as the lines at the top start to darken, the electron beam fills in the spaces between the darkening lines with bright new information. As a result, the overall picture maintains an even brightness. So, the interlace scanning system is the method used to ensure that picture has an even brightness throughout instead of having separate bright and dark bands.

With today's technology it is possible to design systems that scan lines in sequence one after the other (line numbers 1, 2, 3, 4, 5, 6, . . .). This type of scanning is called *progressive scanning*. In fact, engineers would prefer to do this because it eliminates some problems created by interlace scanning. However, all current TV sets would be incompatible with such a system. As a result, we still use interlace scanning. This may change in the future. New technologies to be discussed later in this book may allow TV sets to display either interlace or progressive scanning.

1

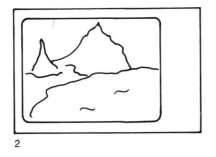

2

1. TV without interlace scanning.

2. TV with interlace scanning.

17

The electron beam needs to get back to its starting point.

Blanking

The intensity of the electron beam during the scanning process is not constant. It varies in a logical, consistent pattern corresponding with the beam's movement across and up and down the CRT's face.

Horizontal blanking

In the scanning process, the electron beam scans the CRT, laying down the first line of information. When the beam reaches the edge of the area defining the TV screen it is turned down to a low voltage, although it continues for a moment in the same direction. When the beam reaches the edge of the CRT, it quickly reverses direction and returns to the other side of the screen. Once it reaches the other side, the beam resumes its original direction. Its voltage is then turned back up to lay down another line. This process occurs every time lines are scanned.

The duration of the lowered voltage from the end of one line to the beginning of the next is called *horizontal blanking*. During this horizontal blanking period, the return of the electron beam from one side of the CRT to the other is called *retrace*.

Vertical blanking

You've seen what happens at the end of each line. Something similar happens at the end of each field. After laying down a field of information, the electron beam is turned down to a low voltage before it retraces back to the top of the image. Once in position at the top of the screen, the beam's voltage is turned back up and it starts scanning a new field. The time that the beam's voltage is turned down until it is turned back up again is called *vertical blanking* or the *vertical interval*. When the electron beam is retracing back to the beginning of a new field, it's called *vertical sync*.

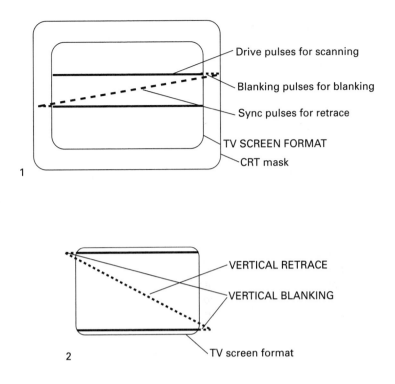

1. Parts of the TV picture.

2. Vertical blanking

Waveform Display

The horizontal scanning movement of the electron beam can be diagrammed another way. Take a look at the figure at the right. This graph represents an electron beam scanning one line of information, returning to the other side of the target during blanking, and reading a new line of video. Everything above the baseline represents the electron beam sweeping across the target in its normal direction, collecting information; anything below the baseline is the beam returning in the other direction.

The graph actually measures the voltage of the electron beam in special units called IRE units (for Institute of Radio Engineers, now called the Institute of Electrical and Electronics Engineers). The baseline is 0 units, which is simply 0 voltage. Notice that the height of the graph ranges from 100 to about –40 IRE. Just above the baseline the black portions of the picture are registered. This area near the baseline where the black portions of the picture appear is called the pedestal.

Above the baseline, 100 IRE units indicate the maximum voltage the video system can handle and still provide a good picture. Information located here is called the *white peak*. Moving from left to right on this diagram, the voltages vary a great deal; as discussed earlier, the higher voltages are the bright parts of the picture and the lower voltages are the darker parts. Then there is a flat line of very little voltage; this is the start of blanking. The blanking continues for a little way and then suddenly drops below the baseline; this is the start of retrace. After retrace, the voltage rises back above the baseline and remains at this low level until a new video line begins.

When you walk around the equipment area of a TV studio, you'll see several displays like the one just discussed. These are waveform displays and they're very important to both production people and engineers. The waveform monitor, which shows these displays, provides a graphic display of the black-and-white portion of the picture. The voltage signal generated from the black-and-white portion of the picture, its brightness, is also called the picture's *luminance*.

The waveform monitor's display will be similar to the simple two-line graph already discussed. Normally, however, the waveform monitor will continuously display the scanning of the electron beam across the target at the 30-frame-per-second rate. The blanking should remain the same from line to line, but the video information will vary as the beam scans higher and higher across the picture. Some monitors allow you to isolate one or two particular lines or points on the graph.

When shown on the waveform monitor, the blanking and retrace are called by different names. The retrace is called the *sync, sync pulse,* or *horizontal sync;* in fact, it's hardly ever called retrace around the studio. Two other parts of the blanking also pick up new names. The first part of the blanking is called the *front porch,* and the last part of the blanking is called the *back porch.*

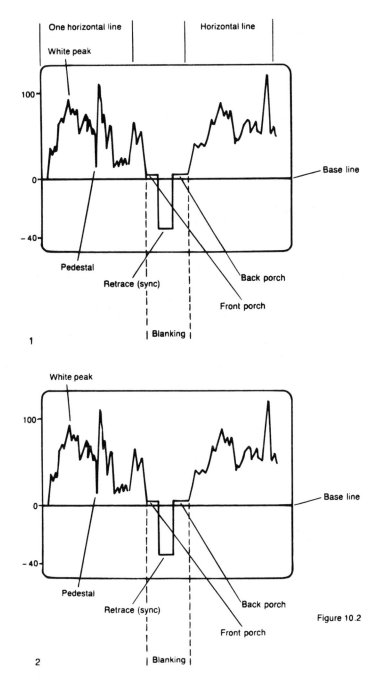

1. Waveform scan.

2. Waveform monitor continuously displaying the scanning of the target by the electron beam.

Charge-Coupled Devices

Charge-coupled devices (CCDs) are the parts inside of the camera that change the light focused on them by the camera lens into electrical signals. CCDs have been developed from the same sort of solid-state silicon chip technology that has made computers faster, smaller, more powerful, and cheaper; put powerful radios into small packages; and made home camcorders the size of a book.

CCD layout and operation

These chips are laid out with horizontal and vertical photosensitive elements (lines). Where the elements cross (called *crosspoints*) are our picture elements (*pixels*). When light hits one of these pixels, a distinct electrical voltage is created. The brighter the light, the higher the voltage; the darker the light, the lower the voltage. All of these discrete voltages are read off left to right and top to bottom into a memory. The memory is then fed out, horizontal line by horizontal line, in sync with the rest of the system. Once the face of the CCD has been cleared, a new image forms and the process is repeated.

Broadcast-quality requirements

To meet broadcast-quality standards, a CCD needs to have around 250,000 pixels. Since having more pixels gives higher detail and resolution, it would be nice to have even more than that. Cramming that many pixels into an area a little larger than your thumbnail had been one of the big problems faced in the past. It took engineers more than 10 years to solve the problems, but they have been very successful and CCD cameras are the norm in most production situations.

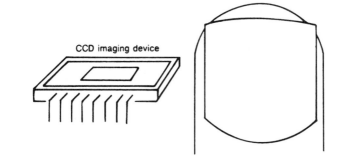

CCD imaging device

1. CCD next to a thumb; the scale is roughly accurate.

2. CCD layout. This diagram shows horizontal and vertical layout of a CCD. Every time a horizontal line crosses a vertical line, a pixel is created.

CCD 1

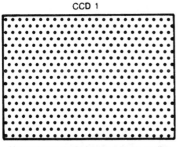

Not enough pixels for broadcast quality

CCD 2

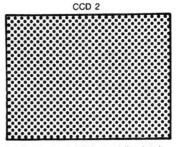

Approaching broadcast quality density

3. Broadcast-quality CCDs need a high pixel density.

4. In the past, CCDs were expensive because each batch produced only a small number of usable chips.

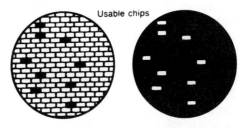

Usable chips

Low % of usable chips   High % of usable chips

23

Using computer technology gives us better pictures.

An Introduction to Digital

What we have been discussing up to now is called analog video. In the analog process, the bright images on the face of the CCD are changed into higher voltages and the darker images on the CCD are turned into lower voltages. These are then reproduced on the CRT where the higher voltages create brighter images and lower voltages create darker images. This process has been around for more than 50 years. But the analog process has a lot of problems. The quality of the picture is limited and can easily be degraded by some equipment. If we could move into the world of computers we could improve the quality of the picture, maintain that quality throughout the process, and create images and effects that can otherwise only be "seen" in the mind's eye. This is what digital video does. Digital video is a process that uses computer technology and language to create, store, and transmit video images.

What computers do

Computers are really very stupid machines. They can only do what we tell them to do and how we tell them to do it. The real brilliance of these machines comes from the men and women who design the internal electronics and who write the software programs that tell the machines how to make those designs work. If you have a basic understanding of what computers deal with, you will have a better appreciation of the people who create and program these machines.

Computers can only do one thing: manipulate numbers. They are number crunchers, calculators. But they can't even handle the numbers 1 through 10 in the same way that you and I do. When you get down to the littlest, tiniest part of the computer's memory it can only deal with one of two things, as discussed next.

Bits and bytes

The smallest part of the computer's memory is called a *bit*. It either has an electrical charge or it doesn't. If the bit has no charge it is represented by the number 0. If the bit has a charge it is represented by the number 1. That's it! That's all computers can deal with: 0s and 1s. As you can see, a 0 or a 1 isn't much information. So in order to have a segment of memory that we can use effectively, we have to combine a number of bits together into a unit that is called a *byte*. In 8-bit processing we would combine 8 bits. In 10-bit processing we would combine 10 bits and so on.

Let's look at 8-bit processing since it is what is used in home computers with which you may be familiar. Since each bit has only two possibilities, 0 or 1, when we combine two bits together we have four possibilities: 00, 01, 10, or 11. When we combine 8 bits together we have 256 possibilities (2^8 or 2 multiplied by itself 8 times: $2\times2\times2\times2\times2\times2\times2\times2=256$). With 10-bit processing you would add 2 more bits to the process and get 1,024 possible combinations. With the 256 combinations of 8-bit processing, we can assign one specific combination to represent the capital letter "A" and a different specific combination to represent the small letter "a." We can do the same thing for the letters "B," "C," "D," and so on. We can also use specific combinations to represent punctuation marks, the numbers 0

1. Bits.   00 01 10 11

                                                          1

2. Bytes (8-bit processing). | 10001101 | 11000101 | 01010111 | 10101101 |

                                      2

through 9, mathematical symbols, and so on. The software, the computer's program, tells the computer how to use that information. "Treat this specific byte combination like it is the letter 'A' in the alphabet." The byte, then, is the smallest segment of memory that can be used as a piece of information: a letter or a digit or a punctuation mark or such.

So no matter what a computer may seem to be doing—surfing the web, writing a paper for class, or creating wild video images—what it is really doing is manipulating numbers (0s and 1s).

25

Analog and Digital

A-to-D conversion

The process of converting information from analog to digital is fairly complex, but it is important that you understand as much of it as you can. Changing the signal from analog voltages into a stream of numbers for digital video takes place in a device know as an *analog-to-digital (A-to-D) converter.*

The video signal, from 0 to 100 on the waveform monitor, is approximately 0.7 V strong. You know that the stronger the voltage (0.7), the brighter the glow on the CRT; the weaker the voltage (0.0), the darker the CRT. If we picked a point right in the middle of that range (0.35 V), we could probably figure out that would be in the middle of the gray range. This is sort of what the A-to-D converter does. The brightness range is divided into 256 possible levels (8-bit processing) with 0.00 V equaling the 0 level and 0.7 V equaling level 255. You may think it strange that we only go up to number 255 when there are supposed to be 256 levels. This is because the first level is 0, the second level is 1, the third level is 2, and so on. As a result, 0 to 255 equals 256 levels. Midway through the voltage range (0.35 V) would also be midway through the number range (127). As a result, each of the 256 numbers represents a separate and distinct voltage and brightness (remember, these voltages have been rounded off).

Sampling and quantizing

The process of grabbing a piece of the video information and holding it is called *sampling.* The process of changing that sample into a number is called *quantizing.* How often a sample is taken is very important. If we sampled the video only a few times a line, we wouldn't get a very accurate picture of what was happening. If you think of the brightness levels along one axis and the sample points along the other axis, you can turn the waveform display into a point-by-point graph that is quite representative. That has been done in the illustration on the right. Imagine how much more accurate it would be with several hundred sampling points instead of a few. The more times we sample a line of video the more accurate will be our digital representation of that line. You could think of the sampling points on the video lines as being individual pixels. For a television picture you will need anywhere from about 700 pixels to almost 2,000 pixels per video line.

D-to-A conversion

To display a picture, the stream of digital information (numbers) must be converted back to a stream of analog voltages. This is necessary so that the electron gun at the back of the CRT can shoot a stream of electrons to spray the picture onto the phosphors on the face of the CRT. This takes place in the digital-to-analog (D-to-A) converter. Since each number in the digital stream represents a specific voltage, the D-to-A converter takes the right amount of voltage from the wall outlet and replaces the number with that voltage. As a result, a stream of numbers enters the D-to-A converter and a stream of voltage leaves it.

26

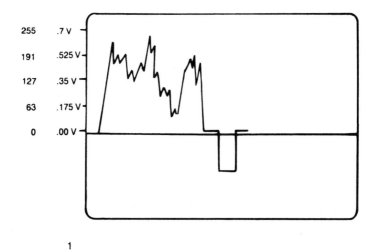

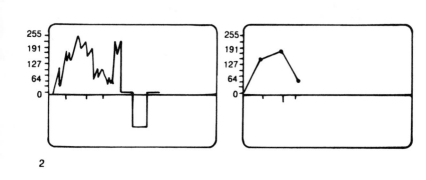

1

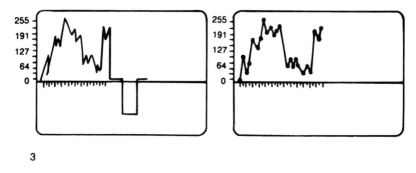

3

1. Breaking the video signal down into numbers.

2. Sampling only a few times a line.

3. Sampling several times a line.

Color Systems

Color versus black and white

You've been shown how a visual image is created on a CCD, converted to an analog video signal, and transferred to a CRT for display. Black-and-white video systems operate this way: A black-and-white camera, if one were made today, would have one CCD, and only black-and-white information, that is, luminance, would be created and conveyed to the CRT. Color video is based on the same basic principles, but is a little more complex.

Additive and subtractive colors

Instead of having just one CCD, high-quality color cameras have three CCDs, one for each of the primary colors: red, green, and blue. Those of you who have had art classes might say, "Now wait a minute, the primary colors are red, blue, and yellow." Well, you're right, if you're dealing with subtractive colors. Subtractive colors are what you deal with if you're mixing paints; subtractive colors reflect light off themselves. In video, we're dealing with additive colors. Unlike subtractive colors, which depend on substances that interact with white light, additive colors depend on the color of the light itself. With additive colors, the primaries are red, green, and blue.

Complementary colors

You can see by the figure at the right that by mixing what appears to be equal parts of any two primary additive colors you will get one of the complementary colors; that is, equal parts of red and green will produce yellow, red and blue will produce magenta, and blue and green will produce cyan. When all three primary colors are mixed together we have white light. By changing the proportions of the mix, brightness (intensity), and saturation of the colors, an infinite number of colors in the visible spectrum can be produced.

The beam splitter

After entering the camera through the lens, the light goes into a beam splitter. Using prisms and/or special mirrors the beam splitter separates the light into the three primary colors. Only light in the red spectrum reaches the red CCD. Light in the green spectrum reaches the green CCD. Light in the blue spectrum reaches the blue CCD.

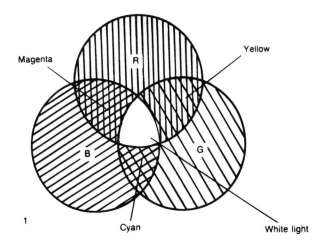

1

Magenta

Yellow

R

B

G

Cyan

White light

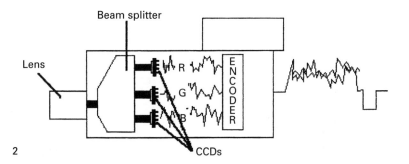

Beam splitter

Lens

R

G

B

ENCODER

2

CCDs

1. TV colors.

2. Color TV camera.

How the Eye Sees Light

In the section on complementary colors it was indicated that a mix of equal parts of red, blue, and green light would produce white light. Yet, in reality white light is 59% green, 30% red, and 11% blue. How is this contradiction explained? It isn't really a contradiction. Your eyes are not equally sensitive to all colors of light. What looks like equal amounts of red, green, and blue aren't really equal amounts at all. The camera's CCDs, however, are equally sensitive to all colors of light. For your eyes to see white on the television screen, the TV camera must produce a picture that is 59% green, 30% red, and 11% blue.

Color temperature
If you've ever taken photographs outside and then gone indoors and taken more pictures without using a flash, you've gotten some disappointing results. Chances are that the pictures taken outside looked great, but those taken indoors had an orange/yellow cast to them. What you've seen is the result of the difference of color temperature. Color temperature is a way of measuring the color characteristics of light. Color temperature has nothing to do with heat or cold. The higher the color temperature, the more blue there is in the light. The lower the color temperature, the more orange is in the light. The light outside on a bright, sunny day might be around 5,000 K (Kelvin, the scale used to measure color temperature). The light in your home is probably around 2,600 K. This is why the colors in the photographs look so different. If you use a flash indoors, the light will be the same color temperature as daylight and your photographs will look fine.

You don't see the change in color temperature because your eyes and brain compensate for you. If you see a person in a yellow jacket outdoors and then the two of you go indoors, your brain knows that it's the same jacket and it couldn't have changed colors. Your brain sees the jacket as the same color. A camera can't do that.

Filters
What a camera can do is use filters. Most cameras have a built-in filter wheel. This is a device that will allow you to place one of several filters between the lens of the camera and the beam splitter. Most cameras are set up to operate with TV studio lights, which have a color temperature of 3,200 K. If you go outdoors to shoot, you would change the filter wheel to compensate for the change of color temperature.

White balance
Changing the filter wheel is only the first step. Because clouds, shade, reflections, and other conditions all have an effect on color temperature, you will have to white balance your camera every time you set it up in new lighting conditions, which is simply establishing the proper color combination for given lighting conditions. To white balance a camera, simply aim and focus it on a white card and push the white balance button. The electronic circuitry of the

30

Overcast sky	7,000 K	Blue
Noon daylight	5,000 K	
TV studio lights	3,200 K	
Household lights	2,600 K	
Sunrise/sunset	2,500 K	
Candle	1,900 K	Orange

1

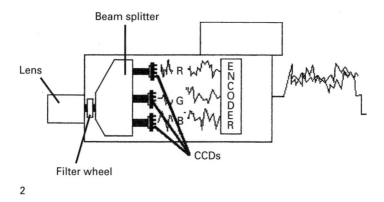

2

1. Approximate color temperature of some common light sources.

2. Color TV camera, showing location of filter wheel.

camera will then adjust its light reception so that the green CCD produces 59% of the picture, the red CCD produces 30% of the picture, and the blue CCD produces 11% of the picture (see previous discussion). Since the camera is now properly mixing the three primary colors, the rest of the color spectrum will be fine.

Digital Encoding Ratios

From black and white to color

When color television was finally adopted in the United States, black-and-white TV had already been around for several years. To ensure that black-and-white TVs and the new color TVs could show all of the same programs, engineers decided to start with the black-and-white signal and add the color information to it. The black-and-white or luminance signal is designated as "Y." Since each of the three primary colors contains luminance information, it would be redundant and use too much bandwidth to have a separate luminance signal and three separate color or chrominance signals. The solution is to take the Y channel for luminance, which is mostly green information (since green makes up 59% of white light), and two color signals: red with the luminance information removed (R − Y) and blue with the luminance information removed (B − Y). This allows recreation of the full color picture without using more bandwidth than necessary.

Digital responses to this situation

When developing a digital system engineers have to make choices about how accurately they want to reproduce the picture and how much bandwidth they want to use. One of the choices they make is with digital encoding ratios. A digital encoding ratio is the relationship of how much luminance and how much chrominance information is encoded digitally. The international broadcast standard is 4:2:2. What that means is that if we were to take four pixels from the TV screen all four of them would have the Y or luminance information encoded, and two of the pixels would have the R − Y information and the B − Y information encoded. If we had 4:1:1 encoding, which some companies use, then all four pixels would have the luminance information encoded, and one would have both the R − Y and B − Y information encoded.

Obviously the more information that is encoded, the more accurate is our picture. But you will use more bandwidth and will pay more money for the better quality. A problem is created, however, when you try to use several pieces of digital equipment together. If one item has 4:2:2 encoding and you try to hook it up to something that has 4:1:1 encoding it probably won't work. The equipment may have built-in converters, but when you constantly convert from one digital format to another, you begin to lose quality. It is critical that you and your engineering staff ensure that all digital equipment in a facility is compatible with each other.

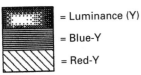

= Luminance (Y)

= Blue-Y

= Red-Y

PIXELS

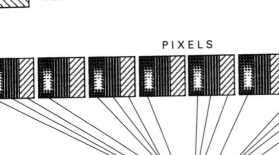

ENCODED

1

ENCODED

2

1. **4:2:2 Encoding.**

2. **4:1:1 Encoding.**

Encoding combines many signals into one.

Composite Encoding

The three color signals are combined in a process called *encoding*. The encoded signal, now called a *composite signal*, actually comprises two parts: the chroma (color) signal and the luminance (brightness) signal. Although there is some variation, the separate luminance signal is formed by skimming brightness information from each of the three CCDs. It is then recombined with the chroma signal to create the composite signal. If you had only a black-and-white TV, you would receive and watch only the luminance signal. The composite signal, then, is one in which all of the components of the picture (luminance and chroma) are combined into a single signal. The encoded, composite signal is fed out the back of the camera.

Home video cameras

Note that some color home video cameras use only one chip instead of three. These chips have a filter of colored stripes in front of them. This striped filter breaks up the light into the three primary colors. Circuitry in the camera then combines these stripes of color into a composite color picture. As a result of this process, these cameras seldom approach broadcast quality. They rarely have the picture sharpness or color saturation needed for professional work. Even home video cameras that have three chips are of poorer quality than professional cameras. The chips that they use have fewer pixels and are not of the same quality as those used in professional cameras. Therefore, these cameras are of little relevance to a discussion of broadcast television.

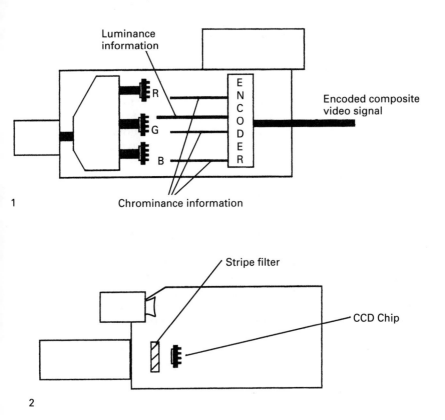

Luminance information

Encoded composite video signal

Chrominance information

R

G

B

ENCODER

1

Stripe filter

CCD Chip

2

1. Encoding.

2. Color home video camera.

35

Color Cathode-Ray Tubes

Just as the color camera works in a manner similar to the black-and-white camera, the color CRT works in a manner similar to the black-and-white CRT. The color CRT has three sets of color phosphors laid down on the inside of the CRT. These are laid down in a specific pattern. The exact pattern differs from manufacturer to manufacturer, but for the sake of explanation, a very common pattern, the triad, will be used. As you can see from the first figure at right, the triad is made up of one red phosphor dot, one green phosphor dot, and one blue phosphor dot arranged in a triangular pattern.

At the back of the CRT there are three separate electron guns, one each for the red, green, and blue information. Close to the front of the tube is a thin metal mask. This mask has tiny holes in it, arranged in such a way that only the red electron beam can strike a red phosphor, the green beam a green phosphor, and the blue beam a blue phosphor. These phosphors are so small and close together that they cannot be seen as separate and distinct unless they are looked at under magnification. Thus, when they are struck by the electron beams, their colors blend to produce the same color that the camera split up. The red part of a stop sign, for example, would cause only the corresponding red phosphors to be illuminated, while the white letters of the sign would cause all of those corresponding phosphors to glow.

Convergence

CRT can experience a problem called *convergence.* If you look at the diagram at the right you will see that each of the electron guns at the back of the CRT is a different distance from the phosphors. The gun that is mounted at the lowest place in the CRT neck will be farther from the phosphors at the top of the screen than the gun that is mounted at the top of the neck of the CRT. Since the electrons travel at a constant speed, the electrons that have to travel a longer distance will arrive at the phosphors an instant later than electrons from the other guns. This is the convergence problem.

If there is a convergence problem, we could see up to three distinct offset images of different colors. Convergence shouldn't be a problem with most home TVs, but it can be a problem with some large-screen projection units. You can usually make some control adjustments to lessen or minimize convergence problems.

We can see from this discussion that the color video system is considerably more complex than a black-and-white system. In fact, the color camera might be thought of as three synchronized cameras and the color CRT as three separate CRTs synchronized together.

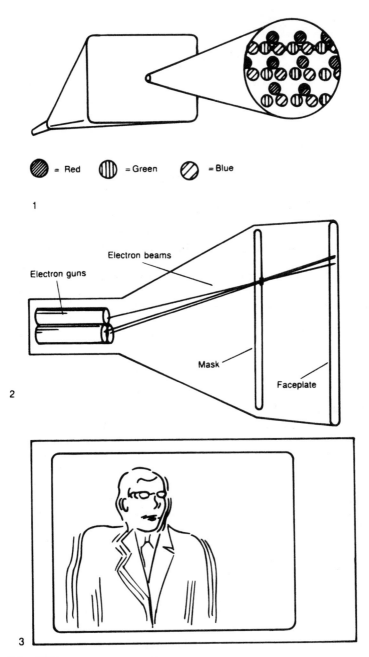

= Red   = Green   = Blue

1

Electron beams

Electron guns

Mask

Faceplate

2

3

1. Color CRT.

2. Components of a color CRT.

3. CRT convergence error.

37

Plasma Display Screen

There is a limit to how large you can make a CRT. As the CRT gets larger and larger, the glass enclosure must get thicker and thicker to maintain strength. A new kind of screen is available for showing video that gives a larger viewing area without the size and weight problems of very large CRTs. The device is only about 4 to 5 inches thick and is sometimes called a *flat screen* display although the proper name is *plasma display screen.*

The plasma display screen is made up of two pieces of glass sandwiched together. The back part has ridges going vertically down the glass. In the space between the ridges alternating columns of red, green, and blue phosphors are laid down. For each color of each pixel there is an electrode on the back that is called the *data electrode.* The front piece of glass has horizontal ridges across that will seal off each pixel from the others. The front panel also has two transparent electrodes for each pixel color; a scan electrode and a common electrode. When the front and back panels are sandwiched together rare gases (helium, neon, and xenon) are trapped into the pixel compartments. So if we were to look at an individual pixel of the plasma display screen there would be a red compartment with a trapped rare gas that has a data electrode on the back and transparent scan and common electrodes on the front. There would be a green compartment with the same setup and a blue compartment with the same setup. The three compartments together would make one pixel.

How it works

To activate a color of a pixel, an electrical charge is sent to the scan and data electrodes on the front and back of the screen. This charge electrifies the rare gas inside the pixel segment. A gas with an electrical charge is called a *plasma,* which is where the name of the screen comes from. The plasma gives off an invisible ultraviolet light that causes the phosphor to glow. The phosphor will continue to glow as long as the charge is held and the plasma is active. To turn off the phosphor, the charge is drained off through the common electrode. When the charge is drained, the plasma returns to its uncharged, nonplasma, inactive gas state and the phosphor stops glowing. By activating individual pixel colors and varying the charge applied to the plasma, a full range of colors can be created.

Besides being only 4 or 5 inches thick, being able to be hung on a wall, and having a large viewing area, these displays have an excellent contrast range and present very bright pictures. At the time of this writing they are still quite expensive. In a few years, however, they should be competitive with other large screen displays.

38

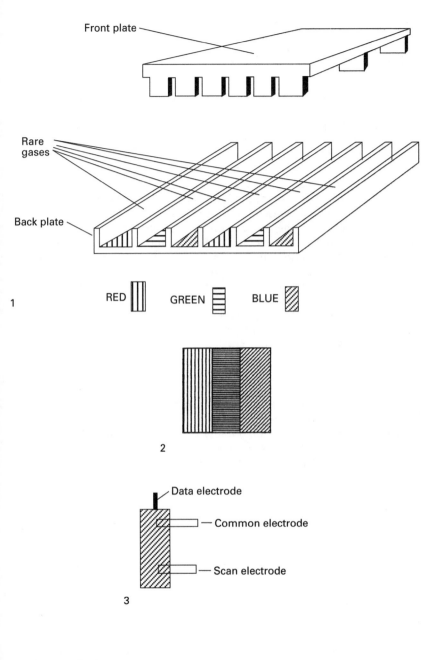

Front plate

Rare gases

Back plate

1

RED   GREEN   BLUE

2

Data electrode

Common electrode

Scan electrode

3

1. **Plasma display screen.**

2. **Individual pixel.**

3. **Pixel element.**

Sync Generators

What exactly is a *sync generator*? A descriptive name would be color synchronizing pulse generating system. No wonder it's simply called a sync generator. The sync generator is the master clock that coordinates the whole system. Like a drummer in a band, it sets the rhythm that keeps everything in time and running together as a unit, rather than as a bunch of individual components, each doing its own thing. When you're working with a single camera, it's not that big a job, but when you start developing a more complex system, the job becomes more difficult.

Imagine that you have two separate cameras working independently of each other. You turn their power on, and off they go. The fact that the CCDs are reading out different parts of the picture is no big deal. But what happens if we try to hook those cameras together through a switcher (a device that allows you to instantaneously change between different video sources)? In the diagram at the right, the electron guns of each CRT are in different locations. Assume that camera 1 is being shown on the program monitor (the monitor that shows what is being recorded or transmitted). What happens if you try to make an instantaneous change (cut) to camera 2? If the cameras are not synchronized, the picture on the program monitor will roll, jump, flicker, or tear—what's called a glitch. In other words, there will be a major picture disruption or breakup when the cut is made.

Imagine, on the other hand, what would happen if the electron guns on the two cameras had been scanning together. Depending on the switcher, there might still be a breakup, but it would be less severe. The sync generator is the remedy to this problem. It keeps everything in the system together so that you can make clean transitions between cameras.

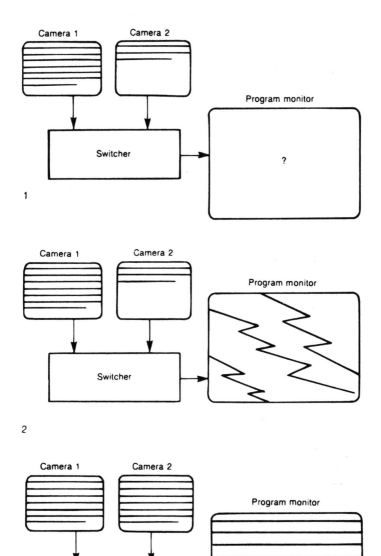

1. An unsynchronized system.

2. Making a transition with an unsynchronized system.

3. Making a transition with a synchronized system.

Sync Generator Signals

Drive pulses

The sync generator provides four primary types of signals, and each of these signals has a couple of subtypes. The first thing the sync generator provides is *drive pulses*. These are the signals that tell the camera what line of the CCD to read off. There are two types of drive pulses, one each to control both the horizontal and vertical position of information being read off of the CCD. They are called horizontal drive pulses and vertical drive pulses.

Blanking pulses

The second type of signal the sync generator provides is *blanking pulses*. As you might suspect, the blanking pulses tell the electron gun when to lower its voltage at the end of a line or field. There are two types of blanking pulses: horizontal blanking pulses (end of a line) and vertical blanking pulses (end of a field).

Sync pulses

The third type of signal the sync generator provides is *sync pulses*. The first two types of sync pulses should be pretty easy for you to figure out. Horizontal sync pulses tell the electron gun to retrace to the beginning of the next line (these are the sync pulses seen below the baseline on the waveform monitor). Vertical sync pulses are at the end of a field and tell the electron gun to retrace to the top of the picture. The third type of sync that the sync generator generates is color sync, also called color subcarrier, color burst, or 3.58 (pronounced "three five eight," the meaning of which is explained next).

Color burst

Color burst is a reference signal that is inserted in the back porch of every line of video. This reference is used when encoding the color information with the luminance information. The color burst acts as a marker, and the color information for each line is encoded onto that line based on the marker. If the marker is off, so is the color. The key aspect of that marker is its frequency. The color burst is really a burst of energy at a specific frequency. That frequency is 3,579,545 Hz ± 5 Hz. This is normally rounded up to 3.58 MHz, hence the name 3.58. If that frequency is off, the colors will be off. This is, then, an extremely important signal that the sync generator provides in a color system.

Combining sync with video

Take a moment and think about how the video system has been described here. The CCDs in the cameras are converting the light images into electrical signals, but the drive, blanking, and sync pulses that tell those CCDs to read out a line, turn down, retrace, turn back up, read out another line, start a new field, and so on are coming from outside the camera. Yet, while coming out of the camera, the video and sync information are combined. This is called *composite video*. Video information without the sync information is called *noncomposite video*. So when you look at the waveform monitor, you're looking at composite video.

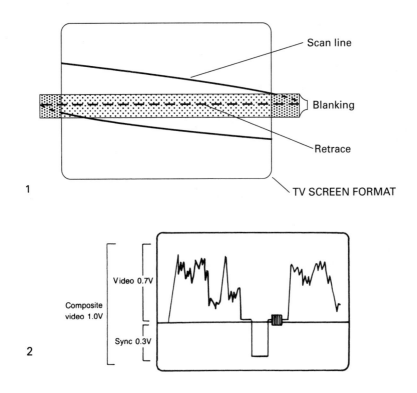

1

Scan line

Blanking

Retrace

TV SCREEN FORMAT

2

Video 0.7V

Composite
video 1.0V

Sync 0.3V

1. **Blanking and retrace.**

2. **Components of composite video.**

In a broadcast-quality system, video is always 1.0V (140 IRE units) peak
to peak (from the bottom of the sync pulse to the top of the white peaks)
across 75-Ω impedance. By looking at the bottom figure, you can see that
the video portion is 0.7V and the sync portion is 0.3V (these figures are
rounded off; the video portion is actually 0.714V and the sync is 0.286V),
adding up to 1.0V. The entire video signal is less powerful than the energy
from a flashlight battery

43

Vectorscope

While the waveform monitor presents a graphic display of the black-and-white information in a picture (its luminance), a vectorscope presents a graphic display of the color information in a picture (its chrominance), and allows us to adjust color levels.

Reading the vectorscope

Like the waveform display, a vectorscope displays signal patterns for scan lines on a CRT. Both the waveform monitor and vectorscope continuously display 525 lines (one video frame) 30 times every second. Since the phosphors on the faceplate of the vectorscope's CRT maintain an afterglow, we actually see the display of a great number of lines at once, producing a composite of the whole picture. The vector display is round, with both black and white located in the center. The little loop in the signal to the left of center (at the 9 o'clock position) is the color burst. The primary and complementary colors are assigned locations in relation to the color burst. Going in a clockwise direction, yellow is about 10° beyond the color burst, red 76° beyond the color burst, magenta 120° beyond the color burst, blue 190° beyond the color burst, cyan 256° beyond the color burst, and green 300° beyond the color burst. This assignment of information in relation to the color burst determines the color's hue.

In the figure at the top you see the typical vector display. In the figure in the center you see what is really the same display, but it has been rotated 90° to the right. The relationship between the color burst (hard to see behind the heavier lines) and the individual colors is unchanged. If we tried to hook together cameras with these two displays in a system we would have a phase problem, a phenomenon to be explained later.

Color bar display

Color bars are reference signals produced by the sync generator and placed at the beginning of a tape when it is recorded. If we read the color bar display on the vectorscope, we can adjust the video system for proper color. Probably the only time production personnel will need to use a vectorscope is for adjusting color.

The sequence of colors in the color bar display is identical to the path of the signal displayed on the vectorscope: from center (both white and black) to yellow, cyan, green, magenta, red, blue, and back to center.

Look at the vectorscope diagram. You will notice that each color falls into its own little box on the vectorscope. This indicates that you are seeing the full level of chroma. Take the red bar, for example. If the chroma were cut in half, the trace on the vectorscope for the red color would stop about halfway between the center of the display and the red box. The basic color would still be red, but it would be a paler red.

When playing back a videotape, you'll look first at the display of the color bars and, if necessary, make a few adjustments to get the same pattern as in the first figure. Once the color bars are right, the colors in

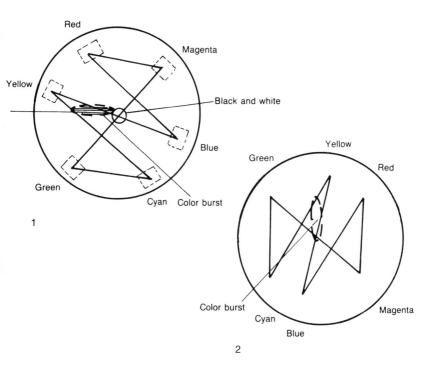

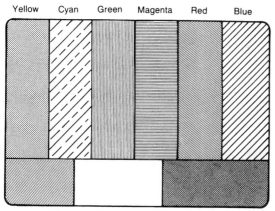

1. **Schematic of a vectorscope monitor.**

2. **Vector display rotated 90 degrees.**

3. **Color bars.**

the video images that follow on the tape should accurately represent the colors in the scene that was originally photographed.

PAL

The system you have been studying up to now is called the NTSC system. It was named for the National Television Systems Committee, which created it in 1952 when the United States adopted its color television system. NTSC is used in North America, Japan, and many other parts of the world.

When the countries of Europe decided to adopt a color TV system, however, NTSC was 10 to 15 years old and its weaknesses and problems were well known. In addition, technology had made long strides that allowed the Europeans to develop television systems that were superior to NTSC. The most commonly used system besides NTSC was developed in Germany and is called *PAL* for phase alternate lines. The luminance and color detail of the NTSC system are not really very good. The PAL system improved luminance detail by increasing the signal bandwidth and the number of scanning lines to 625. The bandwidth is the frequency space that is needed to send a signal through the air or a cable. The wider the bandwidth, the more information can be sent. The frame rate for PAL was, however, dropped to 50 fields and 25 frames per second to match Europe's electrical frequency of 50 Hz.

Except for some minor details, the color signal principles for PAL are the same as those for NTSC. One of the problems of NTSC, however, is that recording and signal transmission can induce errors in the relationship between the color burst and the color information. The PAL process averages out and cancels those errors. In PAL, the phase of the color signal is reversed by 180° from line to line (thus the term phase alternate lines). As a result, PAL television sets always reproduce the correct hue and do not have tint or hue controls, as NTSC television sets commonly have.

PAL systems are used predominantly in continental Europe, Russia, the United Kingdom, China, Australia, and some African countries. A 30-frame, 525-line version of PAL, called PAL-M, is used in Brazil.

The PAL and NTSC vectorscope displays look very similar. Since each line of video in the PAL system uses a color burst that is out of phase with the color burst displayed on the previous line, the vectorscope has to display both. That is the only obvious difference in the PAL vector display.

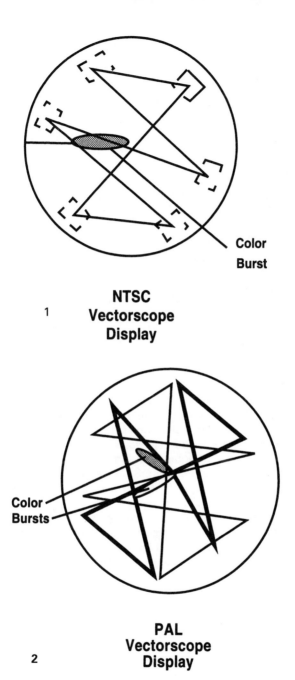

**NTSC
Vectorscope
Display**

Color
Burst

1

**PAL
Vectorscope
Display**

Color
Bursts

2

1. NTSC vectorscope display.

2. PAL vectorscope display.

47

Sync Flow Diagrams (1)

Where do the various sync, drive, and blanking pulses go? The easiest way to illustrate this is by using flow diagrams. Flow diagrams are very handy things to be able to read. They use geometric shapes to represent various pieces of equipment, and lines to represent the wires that connect the equipment. Flow diagrams enable you to see how the various components of a video system are integrated electronically. Most studios are designed using flow diagrams since, if something needs to be changed, it's much easier to erase something on a diagram and draw a new line before a studio is built than it is to rip out a wall to add a new cable after the studio has been finished.

Take a look at the sync flow diagram. This diagram is an electrical map of a typical large (three-camera) studio video system. It charts the electrical pulses going into the cameras, as opposed to the camera flow diagrams (discussed later) that trace the electrical flow leaving the cameras. In the lower right portion of the diagram is a box labeled "Sync Generator/ Video Test Set." This represents our sync generator. Across the top of the sync generator are the outputs: SYNC = horizontal and vertical sync, BLNK = horizontal and vertical blanking, S.C. = subcarrier (color burst, 3.58, etc.), H.DR. = horizontal drive, and V.DR. = vertical drive.

Distribution amplifiers

Follow the path of the vertical drive pulses. You'll see that the path goes straight up and then makes a sharp left turn and comes to a triangle that is labeled PDA 1. That stands for pulse distribution amplifier. A distribution amplifier is a piece of equipment that takes an input and gives you multiple outputs of that same input. So if we have a vertical drive pulse coming into the pulse DA, our diagram shows that we have six of the same vertical drive pulses coming out of it. This is very handy since, unlike home video and stereo, you can't use splitters and other such items because they degrade the signal too much. As its name implies, the distribution amplifier avoids this problem partly by amplifying the multiple outputs it produces. The first output goes to camera 1 to coordinate the readout of video information from the CCDs. The second goes to camera 2 for the same purpose. Similarly, the third one goes to camera 3 for the same reason.

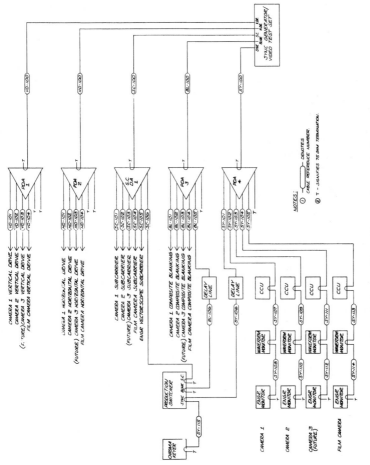

A typical sync flow diagram for a studio system.

49

Sync Flow Diagrams (2)

Termination

The fifth and sixth DA outputs have a T after them. This means they are terminated. But what does that mean? The circuitry of the DA is designed to match up with six pieces of equipment. But PDA 1 is only hooked up to four pieces of equipment; the other two outputs run up against a brick wall, so to speak. This could affect the four outputs that are hooked up. Some of the signal that was supposed to go to the last two outputs could bleed off to the first four outputs. As a result, the first four outputs could be changed from what they are supposed to be. To avoid this problem, a terminator is placed on each of the last two outputs. This makes the circuitry of the PDA "think" that there are pieces of equipment hooked up to it. Since all broadcast-quality video equipment has an impedance of 75 Ω, the terminator is little more than a 75-Ω resistor hooked up to each of the final outputs. Many pieces of video equipment need to be terminated in this way. (In most newer DAs, however, there is enough isolation between outputs so that terminating unused outputs is not required.)

Subcarrier, blanking, and sync pulses

Going back to the sync generator and following the path of the horizontal drive pulses, you can see that they follow exactly the same type of path as the vertical drive. When you follow the path of the subcarrier, however, you see some differences. First, our DA is an SCDA. That simply stands for subcarrier distribution amplifier. The first four outputs of the DA go to the same places as in the two previous examples, but the fifth output goes to the vectorscope. If you think about it, this makes sense. Since the vectorscope displays color information, can't it make use of the color reference? Of course it can. You'll see that the last output of the SCDA goes to the production switcher. The production switcher can modify and do special effects with color, so it needs to have color burst to accomplish this.

The blanking output of the sync generator follows the same path as the other outputs, except that the fifth output goes to the production switcher. That's because the switcher has to have the blanking information in order to do its job properly. Ignore the delay line in that path for the moment; this will be discussed later.

The sync output of the sync generator follows the same path as the blanking did, but it's diagrammed differently. In this case, the first output of the DA goes to the switcher, but this is not of great importance. Taking a closer look at the second output of the PDA, you see that it goes to the camera 1 CCU, then from the CCU to a waveform monitor, and from the waveform monitor to an engineering monitor (labeled ENGR MONITOR). The waveform and engineering monitors are used by the video engineer (or shader) to control the quality of the picture coming out of the cameras. The third, fourth, and fifth outputs of the DA follow a similar path. What

50

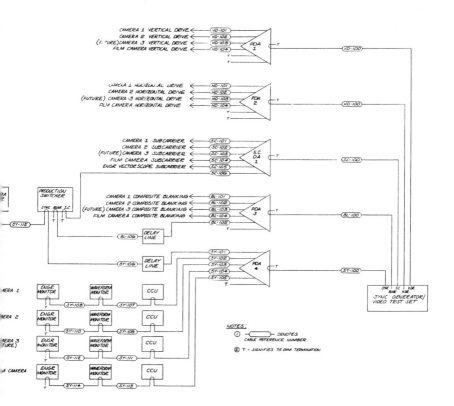

A typical sync flow diagram for a studio system.

these diagrams don't show is that the sync pulses also go from the CCU to the camera head, where they're used for the scanning processes. In addition, all the other sync generator outputs follow the same path: from the DA, to camera CCU, to camera head and waveform and engineering monitors.

51

Camera Flow Diagrams

Take a look now at the flow diagrams for the cameras. As previously mentioned, this diagram traces output from the cameras to their ultimate destinations. The camera head is labeled STUDIO CAMERA 1 and its output goes to its CCU. The CCU has outputs labeled PGM (program), PIX (picture), and MON (monitor). The PGM output is a fully encoded color output and goes from the CCU to a video distribution amplifier (VDA). The first VDA output goes to the monitor bridge. These are the monitors that the director looks at to decide what picture to put on the air. The second VDA output goes to the engineering (ENGR) switcher. This switcher allows the engineer to look at the various video sources to check for problems and fix them before they get too serious. The third VDA output goes to the production switcher. This allows the technical director to record the picture that the director has selected. The fourth and fifth VDA outputs are terminated, and the final VDA output goes to the patch panel (the little circle with the number inside of it). Patch panels are like switchboards; they tie all of the system components together. Patch panels will be discussed later.

The PIX output allows you to look at the red, green, or blue channels independently or in any combination. You can see that this output goes to the waveform monitor, channel A. The waveform monitor has a switch that allows you to flip between two different inputs, and channel A is one of them. As you recall from the sync flow diagram, the output of the waveform monitor goes to an engineering monitor.

The MON output is the same signal as the PGM output. The MON output goes to channel B of the waveform monitor. So by flipping between channels A and B on the waveform monitor, you can compare the complete encoded signal (MON) with the outputs of the separate pickup tubes (PIX).

Camera 1 also has separate red, green, and blue outputs that go to the chroma keyer. The chroma keyer is special circuitry in the switcher that allows for some rather startling special effects. It will be discussed in more detail later.

Dropping down to camera 2, you see that it is just the same as camera 1, except that there are no outputs to the chroma keyer (camera 1 is our primary camera). Camera 3 is the same as 2.

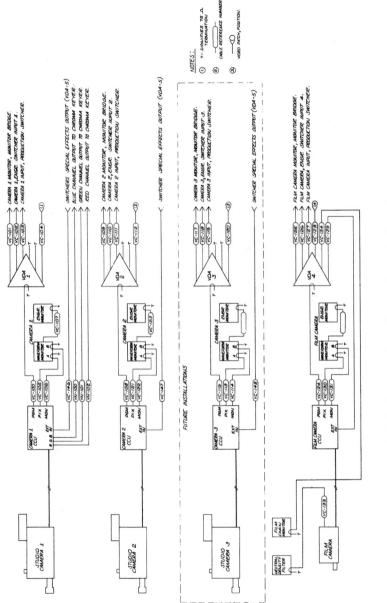

A typical camera flow diagram for a studio system.

NOTES:
① T= SIGNIFIES 75 Ω TERMINATION
② CABLE REFERENCE NUMBER
③ VIDEO PATCH, POSITION.

CAMERA 1 MONITOR, MONITOR BRIDGE.
CAMERA 1,EXAGE. SWITCHER INPUT 1.
CAMERA 1 INPUT, PRODUCTION SWITCHER.

SWITCHERS SPECIAL EFFECTS OUTPUT (VDA-S)
BLUE CHANNEL OUTPUT TO CHROMA KEYER.
GREEN CHANNEL OUTPUT TO CHROMA KEYER.
RED CHANNEL OUTPUT TO CHROMA KEYER.

CAMERA 2 MONITOR, MONITOR BRIDGE.
CAMERA 2,EXAGE. SWITCHER INPUT 2.
CAMERA 2 INPUT, PRODUCTION SWITCHER.

SWITCHER SPECIAL EFFECTS OUTPUT (VDA-S)

CAMERA 3 MONITOR, MONITOR BRIDGE.
CAMERA 3,EXAGE. SWITCHER INPUT 3.
CAMERA 3 INPUT, PRODUCTION SWITCHER.

SWITCHER SPECIAL EFFECTS OUTPUT (VDA-S)

FILM CAMERA MONITOR, MONITOR BRIDGE.
FILM CAMERA, EXAGE. SWITCHER INPUT 4.
FILM CAMERA INPUT, PRODUCTION SWITCHER.

FUTURE INSTALLATIONS

53

Combining Sync and Camera Flow Diagrams

You can combine aspects of the two flow diagrams by eliminating a few parts and combining others to make it simpler. Follow the route of one of the sync signals; all of them follow similar routes. Following the color burst in the figure at the right, you see that the sync generator goes to a DA. The first DA output goes directly to the production switcher. The second DA output goes to the camera 1 CCU. From the CCU, it goes to the camera head and is combined with the video information. That composite video leaves the camera head, returns to the CCU, and from there goes to the switcher. The third DA output follows a similar path: DA to CCU, to camera head, back to CCU, and to the switcher.

Although only one specific system has been discussed, all systems are basically alike, whether using 2 cameras or 20. There will be differences, but the basic technique of running the sync through DAs to the various cameras and taking the composite video from the cameras through DAs to their various destinations will be similar in any professional situation.

Out-of-phase cameras

If you study the figure for a moment, you'll see that each of the paths is a different length. Since electricity travels at a constant speed, it will take the color bursts a different amount of time to follow each of the paths. Thus color bursts on each of the paths will arrive at the switcher at different times and the cameras will be out of phase. What happens if you're on camera 1 and you want to fade it down while you're fading up camera 2 (this is called a dissolve)? Which color burst will the switcher lock up to? It can only use one at a time.

The switcher will lock up on the color burst from camera 1, but as the dissolve progresses, the colors on camera 2 will look funny since they're referenced to the color burst of camera 1 instead of their own. When the dissolve is completed and camera 2 is fully "up," the switcher will lock to the color burst of camera 2 and the colors will snap back to normal. Fortunately, there are circuits in the cameras that can be adjusted to compensate for the difference in distances traveled by the color bursts. However, if the cameras are not set properly and are out of phase, the problem described above will result.

Timing the system

There can be problems with the sync and blanking signals similar to those possible with the color bursts. As is true of color bursts, sync and blanking signals from different sources can arrive at a common destination at different times. That's why there's a delay line between the DAs and the switcher to compensate for the problem. Adjusting the system to ensure that the blanking and sync signals from the various sources enter the switcher at the same time is called *timing* the system. It's very important that the system be carefully timed when it's installed.

Many newer systems use a different approach. Rather than distributing the various separate blanking and sync signals through individual DAs, a single synchronizing signal is sent. This signal is called *black burst*. Black

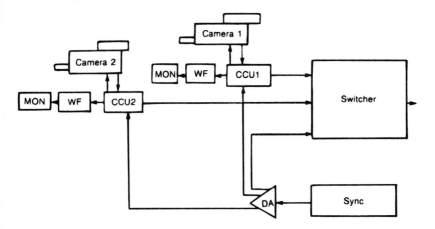

Combining sync and camera flow.

burst originates from the sync generator. It consists of all the normal blanking and sync information along with black video. The individual pieces of video equipment will use the information from the black burst as a sync source. The process of locking up this information is called *gen lock*. This process greatly simplifies the system and reduces expenses.

Switchers allow you to choose among video sources.

Video Switchers

The video switcher is the keystone around which the rest of the TV studio is built. The switcher is a piece of equipment that allows you to choose from many incoming video sources and make transitions or other special effects between those sources. The sophistication of a switcher determines what transitions can be used between shots, what kind of special effects can be used, and how frequently they can be used. The capabilities of various switchers run from the very simple to the mind-boggling, and their prices follow suit. Production switchers can cost from a few thousand dollars to several hundred thousand dollars!

Vertical interval switchers
The standard type of switcher used today is the vertical interval switcher. This switcher has special circuitry that delays any cuts until the entire system is in vertical blanking—the vertical interval. This is one of the reasons why timing the system is so important (see previous page). Since vertical blanking happens 60 times a second, the delay is very small—imperceptible to humans—but it ensures sharp, clean cuts every time. Anything that is going to be switched for use on the air must go through a vertical interval switcher.

Component switchers
The next type of switcher is the component switcher. Rather than using the complete encoded color signal, this switcher deals with the individual red, green, and blue components separately. It's almost like having three separate switchers combined into one package. The three color components travel through the switcher in parallel. This generally produces a much sharper picture and crisper special effects. These switchers do, however, cost a good deal more money than their encoded signal brethren do.

Digital switchers
As the video industry moves more and more toward digital video we also need switchers that are digital. Rather than processing an analog stream of changing voltages, the digital switcher processes a stream of digital information. Just as with analog switchers some digital switchers are designed to handle a single composite digital signal while others will handle separate component digital signals.

Special effects
Virtually all production switchers today come with special effects capabilities. How many of those special effects there are, what they are, how well they work, and how they can be used in sequences all affect the price of the switcher. The most common switcher special effects capabilities will be discussed later.

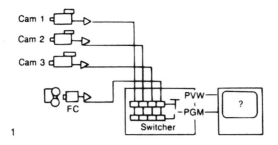

1

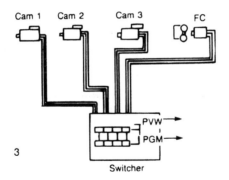

2

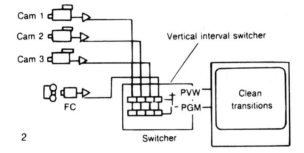

3

1. **Video system using a switcher.**

2. **Vertical interval switcher.**

3. **Component switcher.**

Switcher Applications

Production and editing switchers

A switcher is used in three common situations. In each of these applications you may find either an analog or a digital switcher. In a new facility, you are most likely going to find digital equipment since that is the direction in which the industry is going. The first of these situations is in production. Production here refers to creating finished video that will ultimately be seen by a viewing audience: news, commercials, dramas, comedies, instruction, and just about anything else you can think of. These productions may require anything from the simplest to the most complex production switchers to accomplish the desired results.

On-air switchers

The on-air switcher generally coordinates sources of finished production and sends output directly to the transmitter. It will be switching between various videotape machines, network feeds, satellite feeds, and the studio. These switchers are almost always audio-follows-video switchers. That means that when the technical director pushes a button on the switcher, it changes both the picture and the sound. That's not the case with production switchers, where any changes in sound must be done separately. Since anything that normally goes to the transmitter has the picture and sound together, this makes things much quicker and easier for the technical director. On-air switchers usually have limited special effects capabilities. Because anything going to the transmitter is usually a finished product, there's little need for special effects at this stage.

Routing switchers

The final switcher application is for routing. Say that you're working in a large school system and you need to send seven or eight programs to different classrooms at the same time. You would use a routing switcher to accomplish this task. Routing switchers are often audio-follows-video units.

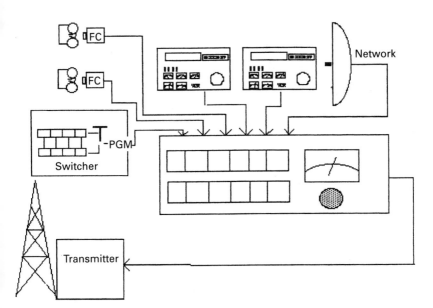

Network

PGM

Switcher

Transmitter

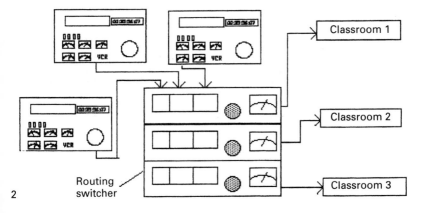

Classroom 1

Classroom 2

Routing
switcher

Classroom 3

2

1. **On-air switcher.**

2. **Routing switcher.**

59

Production Switcher Flow Diagram

Take a look at the flow diagram of a switcher integrated into a simple video system. This switcher is very basic and a lot has been left out to keep things as simple as possible. As you can see, this system has three cameras and one film chain as video sources.

Now follow the outputs of each of the video sources. In each case, the output goes to a distribution amplifier and from there one output goes to a monitor and another goes to the production switcher.

Switcher buses
You'll notice that at the switcher there are two rows of buttons, each row a duplicate of the other. These rows of buttons are called *buses.* They're what give you the ability to cut and dissolve between cameras.

Switcher outputs
The switcher has two outputs. The preview (PVW) output allows the director to see the next shot before it is used. The program (PGM) output is what is intended to be recorded or transmitted. In this example, the final output is both recorded and transmitted.

The preview output is going to a preview monitor for the director's use. The program output is going to a distribution amplifier, and from there one output goes to the program monitor and another one goes to the on-air switcher, where it is fed to the transmitter. The other two program DA outputs are going to videotape recorders (VTRs) where the show is recorded.

This is a simplified diagram of a very basic system, but if you study it, you'll get a good idea of how the basic components are integrated.

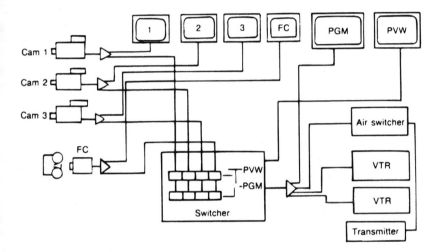

Switcher flow diagram.

Switcher Transitions and Special Effects

All production switchers are capable of at least three transitions: cuts, fades, and dissolves. Cuts are instantaneous changes from one picture to another. A fade-in is a transition that starts with a blank screen (black) that grows progressively brighter until the full picture appears at its normal intensity. A fade-out is the opposite, beginning with a full picture that decreases in intensity to a blank screen. A dissolve is much like a fade, except that as one picture is fading out, another is fading in, so there is always a picture on the screen.

Cuts, fades, and dissolves represent the meat and potatoes of television; they make up most of the transitions used in dramas and comedies. All the other fancy effects that a switcher can do are often called the "bells and whistles." They're there for flash, sparkle, and pizzazz. If your content is solid, you don't need many bells and whistles, although the American public has come to accept them as part of the package. This is not to say that these special effects are valueless. Some can, and often do, add to the content of the program.

Wipes

Wipes are transitions between video sources that are marked by visible edges (sometimes the edge is diffused). Rather than one picture fading out as the other fades in as with a dissolve, in a wipe the new picture replaces the old one by means of a geometric form moving through the old picture. It might be a horizontal or vertical line moving across the picture, or it might be a star or circle coming from the center and expanding until it takes over the whole picture. The number of wipe patterns available seems unlimited. Some switchers come with 40 standard wipe patterns, with still others available as options. Wipes that aren't completed, thus leaving parts of two pictures visible, are called split screens.

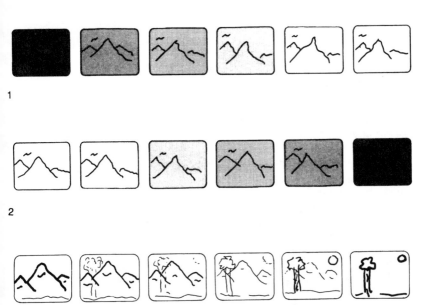

1

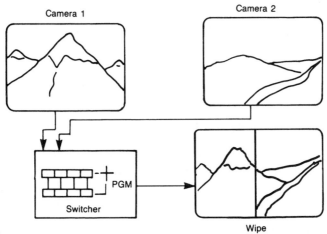

2

3

4

1. **Fade-in.**

2. **Fade-out.**

3. **Dissolve.**

4. **Wipe.**

Special Effects Keys: Luminance Keys

Keys are among the most common and useful special effects. Keys are essentially holes cut into a video picture that are filled with material from another source. Different types of keys rely on different information to determine the shape and nature of the holes and how they are filled.

Luminance keys are triggered by differences in brightness (contrast). The hole in the original video is cut according to contrast differences supplied by a programmed source. The pattern the system derives from this source is called the *key source* or *key signal.* The figure at the right shows a common use of luminance keys.

Camera 2 is providing the key source—the bright star against the black background. The image camera 2 is shooting may come from a card of the black background and white star. The switcher receives the signal from camera 2 and reads the star as a pattern, the key source, based on the difference in brightness between the star and the background. The system cuts a hole in the shape of a star in the camera 1 video.

The hole can be filled in one of several ways. If the pattern that cuts the hole also fills the hole, it is a *self-fill key.* In other words, if the pattern is the white star on the black background, in a self-fill key the hole is filled with white. If the hole is filled by artificially created color from within the switcher itself, it is called a *matte key.* For example, using a matte key, the star-shaped hole in camera 1 could be filled with blue, even if the original key source was black and white. The hole can also be filled with a third source—video from a third camera, for example.

The greater the key source contrast between the desired pattern and its background, the more easily the key operates. This is particularly true when you are shooting a card as a key source, as opposed to using a programmed pattern. White on black is ideal, but for key sources that don't have as much contrast, there will always be a "clip" adjustment that will allow you to compensate. For example, if you are using a black card with a yellow star on it to operate a self-fill key, it will probably work—you'll get yellow color in your video hole—it just might not work as well as white on black.

Keys don't have to be geometric patterns. They can be (and often are) white lettering on a black background—titles, for instance. You certainly can key in other shapes, such as a white horse running against a dark background.

Camera 1

Program

Camera 2

Luminance key.

Special Effects Keys: Chroma Keys

Chroma keys are similar to luminance keys in that a hole is cut out of video, but unlike luminance keys, the triggering device is not contrast. It's a particular color in the subject video (the video receiving the special effect). In a chroma key, the system detects the chosen color in the subject video, and wherever it sees that color replaces it with information from another video source. In the first figure the person (or talent) is in front of a chroma key window. The window can be any color, but blue and sometimes green are used most often. The camera that shoots the chroma key window is called the *source camera.* Any other video source can supply the fill video, but in this example camera 2 has been used. Where the system sees the selected color (the window), it replaces that information with the corresponding information from the fill camera. Thus you get a rocket launch in the studio. If you were to make the entire background the chroma key window, then that background would be filled. The second figure illustrates this. You now see the entire area behind the talent filled with the launch because the background has become the key window.

The chroma key circuitry will try to fill in the source anywhere it sees the chosen color. It is for this reason that primary colors are used for the key window. If you chose yellow for your key window, for example, the system would not only lock up to yellow, but it would try to lock up to anything containing yellow's components, red and green. Thus, the chroma key would try to lock up to almost everything but the color blue.

Since people almost always appear in the chroma key source picture, red is not used for the key window very often. After all, there is quite a bit of red in the flesh tones of people. Although red chroma key windows with people in front of them have been used, it's not a common practice. Blue is the color most commonly used for chroma key windows, with green used occasionally.

Note that talent wearing clothing that is the same color as the chroma key window will present problems, since the clothing might be keyed out as well. Two other colors can create problems with chroma keys when they appear on the key source. Since white includes all colors of light, the chroma keyer may try to lock up to it, but this will usually appear as an incomplete key. Although black is the lack of colors, it too can create problems. Small patches of black, particularly shadows, may also be keyed over. When the system is scanning and it comes to a void of color (black), it may not be sure what to do so it keys over the area. Larger areas of black may not cause a problem because they're large enough for the system to determine exactly what they are.

Just as with the luminance key, the chroma key also has a clip control. In addition, there is a hue (chroma) control that allows you to choose the color to be keyed out, and a gain control that controls the strength of the source picture.

66

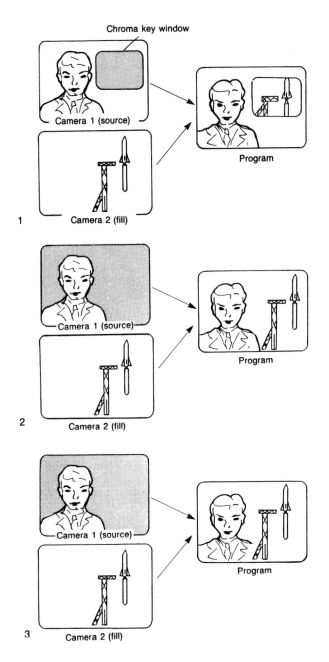

1. Small chroma key window.

2. Full background chroma key window.

3. Talent wearing coat same color as key window.

A studio usually has one of several video signal systems.

Composite Versus Component Video

Problems of composite video

In a previous section you learned that encoding is the process of combining the three chrominance channels and the luminance channel into a single composite video signal. Many pieces of video equipment need to decode this composite signal into its component parts in order to process and use it. Of course, the component signals must then be re-encoded before the composite signal can be sent on its way from that piece of equipment. All of this decoding and re-encoding creates problems. Every time the signal is encoded or decoded, it is distorted a little and a little noise is added. The system has worked well for the last 40 years or so, but now as creative people want to use more sophisticated production techniques and consumers want a better picture, demands are being made for better quality video. There are a number of approaches to the problem.

Component video

In the section on switchers you learned about component switchers that have separate red, green, and blue (RGB) inputs for each channel. Taking the separate RGB outputs from each camera and running them directly to the switcher bypasses the entire encoding circuitry of the camera. This provides a cleaner, sharper picture. Keys come across especially well with this system. Of course, to get the most out of this system the entire studio needs to be component oriented—cameras, distribution amplifiers, switchers, and character generators. This will probably require reengineering of the entire control room since three cables will now have to be installed for every one that was there before. Timing the system becomes even more crucial.

Most equipment that originates video, such as cameras and character generators, is easily integrated into a component system. This equipment already generates RGB signals, so it doesn't take much to bypass the encoder. The DAs and switchers, however, are much more expensive. As a result, it is going to take a big budget to convert the entire studio to component video.

Y/C

A less expensive component system that produces higher quality video than composite video is Y/C. In technical terms, Y stands for the luminance or black-and-white information, and C stands for the chrominance or color information. Use of a Y/C system eliminates many of the encoding and decoding problems inherent in composite video.

To get an idea of how this system works, take a look at Y/C VTRs. Rather than laying down an entire field of composite video in one track, a Y/C must lay down the Y and C information separately to make a field. This provides much better quality at a reasonable cost. Because of these advantages, many feel that Y/C will completely replace composite video in the near future.

68

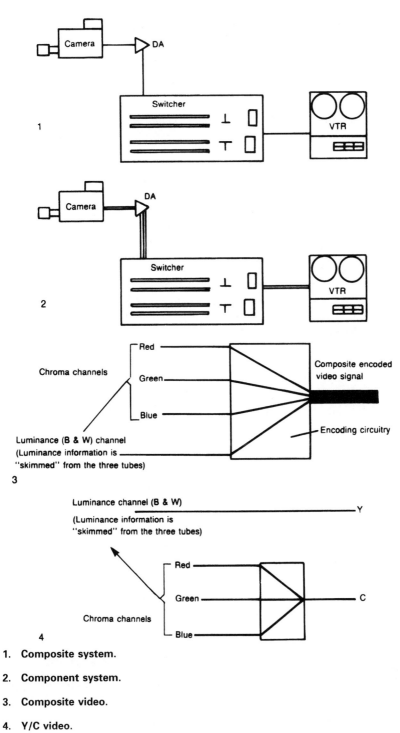

1. **Composite system.**

2. **Component system.**

3. **Composite video.**

4. **Y/C video.**

The possibilities for special effects are endless with digital technology.

Digital Special Effects

Wipes, split screens, luminance, and chroma keys are just some of the special effects available on the typical production switcher. There is another group of special effects called *digital video effects*. Most new switchers made today come with a built-in digital effects system. For older switchers that don't have digital effects a separate stand-alone digital video effects system can be purchased. Each make of digital effects unit will have its own repertoire of effects, often including those discussed next.

Compressions
Compressions are effects that change the entire aspect of the picture: It is made longer, or higher, or the entire picture is compressed into a smaller area. A common use of this is called *chroma key tracking*. With this effect the entire fill video is compressed to fit into the chroma key window. In the first figure, the entire launch has been compressed to fit into the smaller chroma key window. Many stations avoid the chroma key process entirely by compressing what would be the fill video and inserting it into a predetermined space in the program video.

Pushes
Another common digital effect is a *push*. A push, as you might expect, is where one video source pushes another off the screen.

Flips
There are several types of flips. In a page flip, the picture rotates around one edge of the screen as if you were turning the page of a book. Other flips can rotate around a central vertical or horizontal axis.

Rotations
Rotating cubes and spheres are two other digital effects, with video images making up the outer surfaces of these geometric shapes.

Other special effects
The array of possible digital effects is virtually limitless. Video images can be twisted, distorted, curled, and exploded into fragments and miraculously reassembled. In most situations, the limits on the imagination are only restricted by cost and time. Obviously, the more elaborate the effect, the greater the time and money necessary to produce it.

Many digital effects have become practical and affordable due to the development of programmable switchers. These switchers can hold in their memories a sequence or combination of special effects and then recall the arrangement on a command from the user.

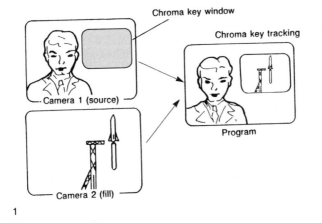

1

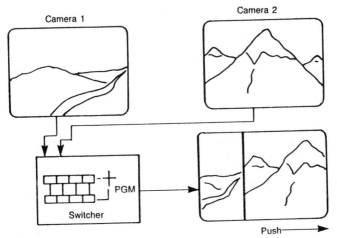

2

1. **Chroma key tracking.**

2. **Pushes.**

A computer is needed to figure out some effects.

Digital Interpolation

As was discussed earlier, digital video is really a series of numbers that represent the brightness and color information for each pixel on the TV screen. Digital video effects are made possible by manipulating those numbers or by creating new numbers through a mathematical process called *interpolation*.

Manipulation
Let's assume that you wanted to compress video into a smaller space such as was described on the previous page in the chroma key tracking example. You want to take the same information and push it into a smaller space, but the number of pixels on the screen is fixed and cannot be changed. You have to remove some information to fit the available space. Assume that you have one line of video that you want to compress down to the space of a half line. A line of video has about 700 pixels. In the digital stream the brightness of each pixel will be represented by a number between 0 and 255. So in the digital world, a line of video would be a stream of 700 numbers. If we removed every other number in the line we would be left with a value for 350 pixels or a half line of video. The one line of video has been compressed to a half line.

Interpolation
But what if you wanted to take a small part of the picture and expand it to fill the screen? Let's reverse the example above. We want to start with a half line of video (350 pixels) and expand it to a full line (700 pixels). Let's deal with only the digital values for the first 8 pixels of that line. We'll say the digital values for those first eight pixels are 0, 20, 40, 60, 80, 100, 120, and 140. If we want to double the length of the line, we have to expand this information and put an additional pixel between each of the values above: 0, ?, 20, ?, 40, ?, 60, ?, 80, ?, 100, ?, 120, ?, 140. But what value do we give to each of these new pixels? In this example, it is pretty easy to figure out because the brightness of the picture is increasing at a steady rate. There is a difference of 20 for each of the values we have. By taking half of that we know how much to add to the value of each pixel and we get this result: 0, 10, 20, 30, 40, 50, 60, 70, 80, 90, 100, 110, 120, 130, 140. We have expanded a line of video and the new line is consistent with the part of the old line with which we started. This process is called interpolation. Of course, in reality, a computer is actually doing this for you. It should be pointed out that this type of manipulation takes very fast computer processing speeds.

The example given here is pretty simple and there are limits to how far you can expand video. What if the computer has to find the values for 50 or 100 pixels between each of the known pixels? That becomes more of a guess than a calculation and can't be done with any degree of accuracy. So when you see a mystery on TV where they take a very small segment from a security camera video and expand the suspect's face to fill the screen with a clear sharp image, it's only TV make-believe, not reality.

72

74 122 176 207 212 224 250 251 154 106

74 176 212 224 251 106

1

20 40 60 80 100

20 **30** 40 **50** 60 **70** 80 **90** 100 **120**

2

1. A video line compressed to half a video line.

2. A half of a video line expanded to a full line using interpolation.

73

Videotape Recording Technology

Recorders

The development of the videotape recorder (VTR) has been fundamental to the growth of modern communications. Videotape recording enables the practical recording and immediate or future playback of a high-quality video image.

The basic theory behind the recording of video information is the same theory that allows us to record sound on tape. Just as with an audio cassette, video recording imprints magnetic signal patterns onto a specially prepared tape. Two of the key aspects of this process are the tape itself and the recording head.

Videotape

The foundation of videotape is a strong plastic ribbon. On the back of the tape is a slick surface that helps the tape move through a mechanical transport smoothly. On the front side of the tape are metal oxides mixed with a binding compound that secures the oxides to the tape. You might think of the metal oxides as microscopic metal filings. If you pass a magnet close to these oxides, two things happen. First, their physical arrangement will be changed, and second, they will be left with their own, much weaker, magnetic field. The stronger the magnet that arranges these oxide particles, the stronger the induced magnetic field.

Recording heads

To record the desired information, a special electromagnet called a *recording head* has to be used. For video, the head is very small, made of very thin metal (about the thickness of a fingernail). The head is hollow, like a tube. Thin wire coiled around the other side of the head connects the head to the rest of the recorder.

In the second figure at right, the head has been enlarged many, many times. Although the figure shows the head by itself, in reality it would be mounted in a small nonmetallic enclosure. The head and enclosure are often shaped something like a piece of bread. The curved surface is the one through which electrical information is exchanged with tape. As the changing analog voltage (the composite video signal) from the camera electronics is processed and flows through the head, it causes corresponding changes in the electromagnetic field that the video recording head produces. This, of course, leaves varied magnetic fields in the oxides on the tape. This is the basic recording process.

The playback process is just the reverse. The magnetically encoded tape is passed across a video head that has no signal flowing through it from another source (such as a video camera). The magnetic field on the tape induces a signal into the head corresponding to the varied magnetic fields on the tape. Thus you reproduce the same analog signal from the tape that was induced onto the tape from the camera source. Since the tape itself wasn't physically changed during the playback process, the signal information is still there for playback.

74

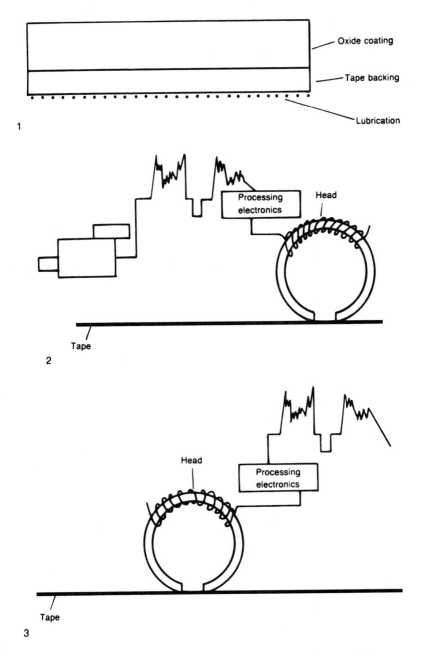

1. Cross-section of magnetic tape.

2. Recording on tape.

3. Videotape playback.

Video Recording Standards and Formats

Audio versus video recording

The recording process is essentially the same for both audio and video recording, but there is about 200 times more information in the video signal than in an audio signal. There is only so much information you can fit onto a given piece of tape. As a result, there's just not enough room for all the video information using audio recording technology.

Thus, there are several significant differences between video and audio recording technology that reflect video's need for greater information capacity. First of all, standard audio tape is ¼ inch wide. Early videotape was much larger (2 inches), although with new technology it is generally ½ to 1 inch wide. Second, standard audio tape speed is 7½ inches per second (ips), while earlier video ran at 15 ips. Finally, audio heads are stationary. Video uses several smaller heads mounted on a rotating disk. All of these characteristics provide video with the capacity it requires to handle the extra information.

Helical video recording

Helical tape machines are the standard of the broadcast industry whether in the analog or digital format. Although digital recording formats are quickly taking over the TV industry, many different formats of analog helical videotape machines can still be found in professional broadcast operations. While none will interchange with the others, they all have some things in common. Although the number of heads mounted on the headwheel varies with different formats, all helical headwheels are mounted at an angle and rotate in almost the opposite direction to the tape's movement. The magnetic heads lay down long slanting video tracks on the tape. Each of these tracks contains one field of video information: video lines, blanking, and sync information. Therefore, if you were to slow down or stop the tape with the headwheel spinning, you would still get 60 fields of information a second, so slow motion and freeze frames are possible with helical machines. However, because of some other problems that will be discussed later, these are not broadcast-quality slow motion or freeze frames.

Other broadcast-quality formats

The different broadcast-quality analog videotape machines use tape that is ½ inch, ¾ inch, or 1 inch wide. These machines are electronically complex, but they produce excellent pictures when you consider that their technology goes back 30 to 40 years. Some of these formats use what looks like a normal consumer videotape cassette. However, the professional formats record at a higher speed and use a special high-quality tape in the cassettes. These are among some of the best analog videotape formats still being used.

76

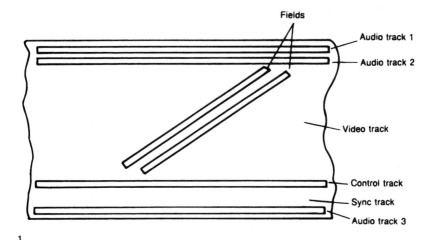

Fields

Audio track 1

Audio track 2

Video track

Control track

Sync track

Audio track 3

1

1. Various videotape tracks (SMPTE Type C).

Sound and control tracks

So far only the video information recorded on tape has been discussed, but a lot of other information is needed on the tape. Sound, for example, also has to be recorded. All of the nonvideo information is recorded using stationary heads much like the heads used in standard audio tape recorders. The number, type, and placement of these other tracks will vary with the specific format of the machine being used.

One of the most important of these other tracks, and one that is common to all formats, is the *control track*. During the recording process, the vertical sync pulses are recorded on the control track. The control track thus helps stabilize the tape's playback speed. You know that the vertical sync pulses are laid down at a rate of 60 pulses a second (one for each field). The regularity of these pulses makes the control track important for other reasons that will be discussed later.

Time Base Error

The tape transport system is a very important and complex mechanism that pulls the tape across the tape heads. It is also the source of a major problem in videotape recording. The problem is that it's impossible to build such a system that operates at a truly constant speed, and that means that it is impossible for the machine to play back tape at precisely the same speed at which it was recorded. You can come pretty close, but in some cases that's not good enough.

If you review the section on sync generators, you'll see that the video signal is composed of some very precise bits of information. Since all this video and sync information is coming from the sync generator and camera, it is being recorded very accurately. But since the VTR can't play back the tape at exactly the same speed at which it was recorded, the playback information won't be as precise as what was recorded. This inability of a VTR to play back at exactly the same speed is called *time base error.* Time base error is measured in lines. It takes the electron gun 63.5 μsec to scan a video line. If the playback signal is 63.5 μsec off from where it should be, there is one line of error.

Dealing with time base error can thus be an integral part of videotape recording. If the amount of error is small or can be corrected, the resulting picture problems will be minimal. However, a large error or, in some cases, any error at all can produce terrible jitters, jumps, and rolls in the picture.

78

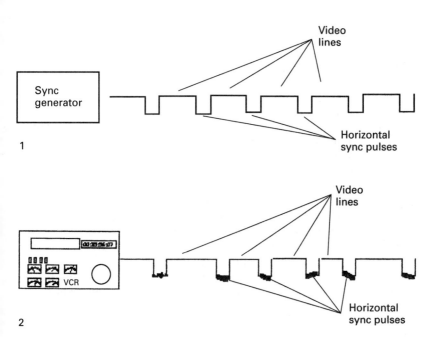

1

2

1. Sync generator puts out very stable signals with each video line being exactly the same length and the horizontal sync pulses falling at the same intervals. This precise signal is then recorded on videotape.

2. Because the videotape machine cannot play back at an absolutely precise speed, the video lines vary somewhat in length and the horizontal sync pulses are reproduced at varying intervals. The degradation of the sync pulses comes from generation drop, not from time base error.

External Causes of Time Base Error

A VTR that spends all of its time in the studio will have relatively little time base error, assuming that the equipment is in good condition and properly used. But VTRs that are used in the field have an additional chance for time base error. Time base error can result from any change in the actual composition of the tape, generally due to variations in air temperature, humidity, recorder position, and recorder movement. When a portable machine is used in the field, it is subjected to constantly changing conditions. It might be in the bright sun one minute, and in the cool shade of a tree the next. You might be shooting at the fog-shrouded seashore in the morning, and in the hot, dry desert later that afternoon. Changes in temperature and humidity will cause the tape to expand and contract. How might this result in time base error?

Assume that you're shooting outside on a hot, sunny day. This will cause the tape to expand. After a hot afternoon of work, you return to the nice air-conditioned studio to edit the piece. The cool temperature of the studio causes the tape to contract. Well, even if the VTR could play back the tape at exactly the same speed as it was recorded, there would still be a problem. Since the tape has contracted, the video tracks have also contracted, so it will take a little less time for the head to scan them. Additional time base error has been created.

Gyroscopic time base error

Gyroscopic time base error refers to the creation of time base problems specifically by changes in recorder position and movement. It's a very common occurrence given today's highly portable video equipment. Let's say that you've got one of those great camcorders that combines the camera and VTR in one unit. You have it up on your shoulder for some great shots and then you swing around to follow the action. *Gyroscopic time base error* has been created. Gyroscopic error occurs because the spinning head drum in the VTR acts like a gyroscope. If you have ever played with a toy gyroscope or a spinning top, you know that if you try to move the toy against the plane of its rotation you will feel resistance. This is what happens inside the VTR. As the VTR is moved against the plane of rotation of the head drum, the spinning drum resists and slows down a little. When it slows down, the information will not be recorded at the proper speed. This is gyroscopic time base error.

All of these factors determine how much time base error you might expect on tape playback. It could be less than a line in a studio machine, or add up to 20 or 30 lines (or more) on a field recorder!

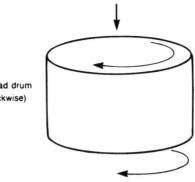

Video recorder head drum
(moving counterclockwise)

A

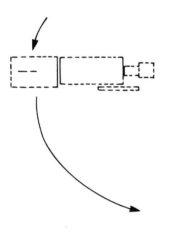

B

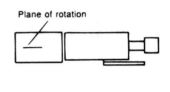

Plane of rotation

C

D

Gyroscopic time base error. (A) Forces that move against the head drum's plane of rotation will cause gyroscopic error. (B) Side view of camcorder showing head drum and rotation. (C) Top view of the same camera showing the head drum's plane of rotation. (D) Gyroscopic error is caused when the head drum's plane of rotation is changed.

A time base corrector can fix time base error.

Time Base Error Correction

Time base error becomes a real problem when you try to integrate video-tape material into a production. If you just want to play back the tape, no problem, but if you want to fade, dissolve, wipe, split screen, or key using taped material, forget it! You've got an imprecise playback signal trying to match up with the very precise, sync generator-controlled video system. If you try to do any of the above-mentioned effects with tape, the picture will jump, jitter, roll, or tear. In short, it will look terrible. This can be corrected with a time base corrector. In fact, if you want to integrate analog videotape into a production, it must be time base corrected. Without time base correction, analog videotape simply doesn't work as a video source in a production. Time base correctors will take the unstable signal coming out of the VTR, stabilize it, and give you a usable signal for production work. Exactly how they work will be discussed later.

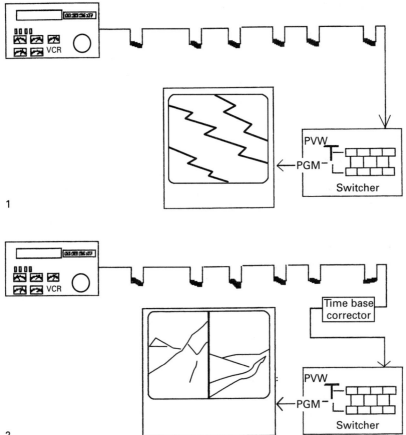

1. Integrating helical VTR video without time base correction.

2. Integrating helical VTR video with time base correction.

VTR Lockup (1)

The more precise we can make the VTR's playback speed, the less time base error will be created. When the machine gets up to full speed and everything is as stable as it is going to get, we say the machine is "locked up." There are several degrees of lockup and each additional step adds a little more stability.

Capstan lock
The first degree of lockup is called *capstan lock*, suitable mainly for home videotape recorders. Capstan lock is not a very stable state. The machine essentially relies on the stability of the power source for a constant base. If the power coming out of the wall varies, so does your tape speed, because the speed control circuitry is very simple.

Vertical lock (capstan servo)
The next level is called *vertical lock* or *capstan servo.* This is the minimum degree of lockup needed to do videotape editing. Capstan servo machines are considerably more complex than capstan lock machines and make use of the control track (remember it?) and incoming sync from the sync generator. If you'll recall, vertical sync pulses are laid down on the control track and, since they come from the sync generator, they are laid down at very precise intervals. A capstan servo machine is also hooked up to the sync generator and has a special circuit to compare the number of incoming vertical sync pulses from the sync generator with the number of pulses being played back from the control track. Since both pulses ultimately originate from the same source (remember that the control track pulses originally came from the sync generator as vertical sync pulses), there should be the same number of pulses in the same amount of time. Since the vertical sync pulses come from the sync generator at precise intervals, they act as a clock. If there are 60 pulses from the sync generator and only 55 from the control track, then the tape is moving too slowly and the capstan servo is signaled to increase the playback speed. But if there are 60 pulses from the sync generator and 63 from the control track, then the tape is moving too fast and the capstan servo is signaled to slow it down.

The big advantage of vertical lock is that a vertical interval switcher will be able to switch to tape from anything without a breakup. However, time base error will still be a problem and dissolves, wipes, and other special effects with tape will not be possible.

The capstan servo machine compares the incoming vertical interval sync pulses to those recorded on the control track to more accurately control the tape's playback speed.

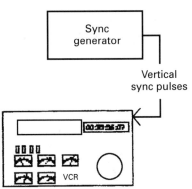

Sync generator

Vertical sync pulses

VCR

VTR Lockup (2)

Frame lock

The next level of lockup is called *frame lock* and follows the same rule of comparing information from the sync generator with playback information to help regulate tape playback speed. With vertical lock, the system was just trying to match one vertical sync pulse from the control track with each vertical sync pulse coming in from the sync generator. There was no attempt to match an even field pulse with an even field pulse or an odd field pulse with an odd field pulse. That's what the frame lock circuitry does. It determines whether the vertical sync pulse coming from the sync generator is for an odd or an even field. Then it speeds up or slows down the tape machine until the pulses off the control track match: odd for odd and even for even. This makes the tape playback speed just a little more precise.

Horizontal lock

We now have the fields matched, but each field has 262.5 lines, and each line has a horizontal sync pulse. This leads to the next level of lockup, horizontal lock. The horizontal lock circuitry compares the number of incoming horizontal sync pulses with the number of played-back horizontal sync pulses. The VTR then speeds up or slows down the tape in an attempt to match the two horizontal pulses.

1. The horizontal lock machine tries to match every horizontal sync pulse played back to a horizontal sync pulse from the sync generator.

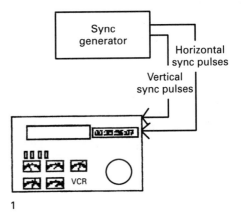

1

A time base corrector turns unstable video into stable video.

Time Base Correctors

Review of time base error problem

In an earlier section you found that it was impossible to make a VTR that would play tape back at the exact speed at which it recorded, and thus time base error is induced into the playback signal. In addition, changes in the environment can create error, and gyroscopic errors can result from the movement of the machine. All of this time base error can create such an unstable signal coming out of the VTR that it can't be integrated into a video production through a switcher.

What a time base corrector does

Total time base error in a typical video can amount to more than 20 or 30 lines and needs to be brought down to the range of 5 to 10 nsec. A time base corrector (TBC) can do this job for us. Unstable video goes from the VTR into the TBC and stable video comes out of the TBC, which is then fed into the system.

How a TBC works

Of course, things are considerably more complex than this. The biggest problem is that we need a good sized memory to store the video picture in, so that it can be fed out at an even, synchronized rate. Memories that will hold a lot of analog video are very expensive and difficult to make. However, computer memories are relatively inexpensive and common, so if we could get our analog video into computer form, we could make good use of those memories. That's just what happens. But, as you learned earlier, the only thing that computers can deal with is numbers. The first thing that happens after the unstable video enters the time base corrector from the videotape machine is that it goes through an analog to digital (A-to-D) converter and is changed to digital information. From there the information goes to a large computer-like memory.

Horizontal sync as a clock

But a method needs to be devised to release this digital video information in sync with the rest of the system. As you might suspect, we go back to the sync generator, which is hooked up to the TBC. The horizontal sync pulses from the sync generator act as a clock. Each time a horizontal sync pulse reaches a special gate circuit, it lets one line of digital video information out of the memory.

D-to-A conversion

This digital video can't be integrated with analog video, so it has to go through a D-to-A (digital-to-analog) converter to be changed back to analog. As a result, corrected analog video comes out of the TBC.

Video proc amp

In addition, TBCs have an internal proc amp (video processing amplifier). As previously noted, the quality of the sync itself may have deteriorated

88

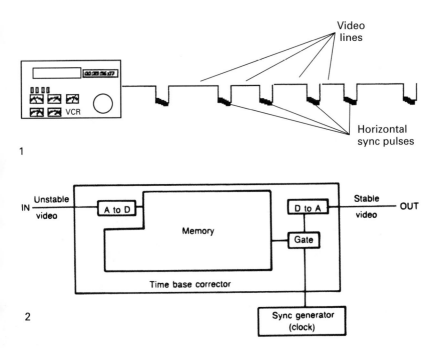

1. Because the videotape machine cannot play back at an absolutely precise speed, the video lines vary somewhat in length and the horizontal sync pulses fall at varying intervals.

2. Time base corrector.

due to factors such as copying a tape. However, proc amps have their own sync generators, so that when the distorted sync comes off the tape, the proc amp strips it away and inserts new, clean sync in its place. Proc amps also enable you to adjust some of the video parameters, such as brightness, chroma intensity, phase, and pedestal (black level).

Window of correction
Each TBC will have what is called a *window of correction.* As you might suspect, this tells you the maximum amount of time base error it can correct. A TBC with a four-line window might be fine for VTRs and tapes that never leave the climate-controlled confines of a studio, but it would be almost useless with tape that is shot out in the field. A TBC with a 32-line window will cost a little more money, but it should be able to handle anything shot in the studio or the field.

Larger Sync Problems and Solutions

Nonsynchronous sources

TBCs are great for correcting the relatively minor errors found on tape, but there are some video sources that are totally out of sync with the studio. For example, when the networks do a football game, do you think there's a cable going from the sync generator on the ground all the way up to the blimp that's getting those dramatic aerial shots? Of course not! Since there is no incoming sync for the blimp camera, it has to be on its own system and it is totally out of sync with the cameras on the ground. We say that the blimp is a *nonsynchronous* or *wild video* source. Or, what about the local TV station? Each station will have its own sync generator, but the video coming into the station from a network or a satellite feed will be on a different sync generator and thus be a nonsynchronous source. If the local station wants to dissolve from the network football game to a local commercial, it could lose sync, which will cause the picture to break up.

Frame synchronizer

The solution to problems posed by nonsynchronous sources is the *frame synchronizer.* Although the frame synchronizer operates a little differently than a TBC, you can think of it as a TBC with a huge memory. It can hold more than a frame of video information in its memory. Like a TBC, a frame synchronizer converts lines of analog video signal into digital form and stabilizes them. However, the vertical sync rather than the horizontal sync acts as the frame synchronizer's gate release signal. When it gets a vertical sync pulse from the house sync generator, the frame synchronizer feeds out a field of video information. As a result, any nonsynchronous video source can be locked up to and integrated with the system you're using. Frame synchronizers are a tremendous resource since they will do anything a TBC will do, plus lock up a nonsynchronous source.

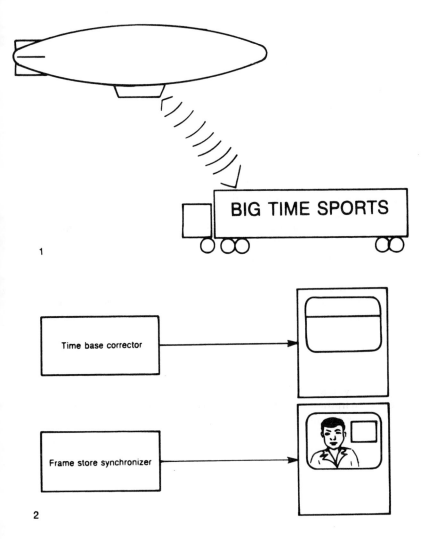

BIG TIME SPORTS

1

Time base corrector

Frame store synchronizer

2

1. Since the blimp and the remote truck are operating on separate sync generators, they are nonsynchronous.

2. A TBC feeds out one line at a time while a frame synchronizer feeds out a field at a time.

91

Slow motion and freeze frames are also possible.

Other Advantages of TBCs and Frame Synchronizers

Dynamic tracking heads

TBCs and frame synchronizers can give you a couple of other capabilities as well. If your VTR has a dynamic tracking head, a TBC will allow you to do broadcast-quality slow motion. A dynamic tracking head is a video head that adjusts itself to follow the video track when the tape changes speed. When you slow down the speed of videotape playback, the relationship between the video head and the angle of the video track changes. The dynamic tracking head automatically compensates for this change and always stays centered on the video track. A TBC or frame synchronizer simply maintains the proper sync with the rest of the system, even if, as in this case, the tape speed is slowed.

Freeze frames

A frame synchronizer will also let you freeze frames. When you push the freeze button on a frame synchronizer, it will continuously feed out the same field of video information. Some units will also feed out a complete frame, but then you often get frame jitter. This is because the video picture is not static. In the camera, after the first field is scanned, the second, interlaced field is scanned. But in the meantime, the subject of the picture may have moved just a little. So when the second field is interlaced with the first one, the subject may be offset a little and this can cause a jitter of the subject between one field and the other.

TBCs, VTRs, and production

TBCs are what have really made VTRs practical for broadcast use. TBCs and frame synchronizers have become an integral part of TV production facilities everywhere. They are the devices that really started the digital revolution in video production.

Digital technology provides another advantage beyond what might be apparent from this discussion of TBCs and frame synchronizers. Unlike analog, when the digital signal is run through several pieces of equipment, there is no increase in noise. Since digital information is a series of numbers, the D-to-A converter will ignore any surrounding noise when it converts that number to its corresponding voltage.

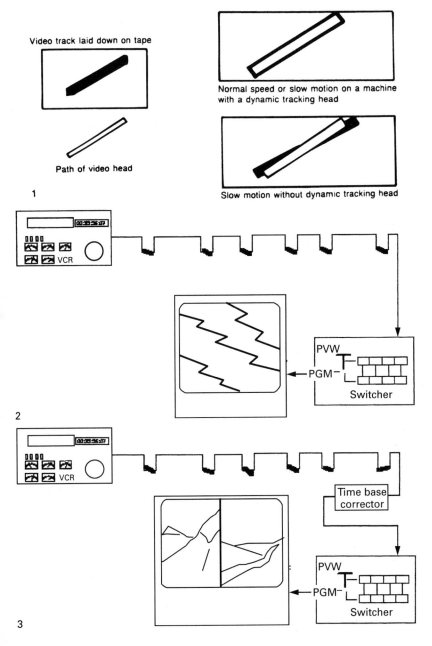

Video track laid down on tape

Path of video head

1

Normal speed or slow motion on a machine with a dynamic tracking head

Slow motion without dynamic tracking head

VCR

PVW
PGM
Switcher

2

VCR

Time base corrector

PVW
PGM
Switcher

3

1. **Dynamic tracking heads.**

2. **Integrating helical VTR video without time base correction.**

3. **Integrating helical VTR video with time base correction.**

93

Editing Videotape

Physical cutting and splicing

In the early days of videotape, editing was a difficult and time-consuming process. A liquid solution had to be put on the tape to make the control track oxide patterns visible. The tape had to be viewed under magnification and then it was physically cut. Finally, the ends were spliced together with a special tape. On top of all this, the roughness that the splice created in the tape's surface often damaged or even destroyed the tape head. Needless to say, editing tape was done only when there was no other alternative.

Electronic editing

Today, however, video editing is done electronically. Information that should be sequential may be out of order on one tape, or it may be distributed among several different tapes. During editing, the material is recorded electronically in the proper sequence onto another tape. The process of copying information from one tape to another is called *dubbing.* Whenever we dub a tape, we drop a generation. The original tape that the video is recorded onto is called the *master* or *first-generation tape.* If we make a copy of that tape, it is a second-generation tape. If a dub is made of the second-generation tape, the resulting tape becomes a third-generation tape, and so on.

Every generation drop also entails a loss of quality. The signal has to go through the entire electronics of a tape recorder, be recorded onto the tape, taken off the tape, and run to another tape recorder. Every step of this process adds additional noise to the signal and diminishes the original quality of the signal. A top-rate VTR can go several generations before this quality loss becomes visible to the human eye. However, a low-quality machine, such as a home VTR will show a very visible quality loss on a second-generation tape. (*Note:* You may refer to your home machine as a VCR, which stands for videocassette recorder. VTR is short for videotape recorder, so VTR applies to all videotape machines, whereas VCR applies to only those machines that have the tape contained in a plastic enclosure.)

Editing allows you to shoot a video production in the most efficient manner, which often means shooting scenes out of sequence from the script. These disconnected scenes can then be rearranged and put together in sequence in the editing suite. The second figure shows the master tape, with the shots in nonsequential order. Through the editing process, the shots or scenes are arranged into their proper sequence.

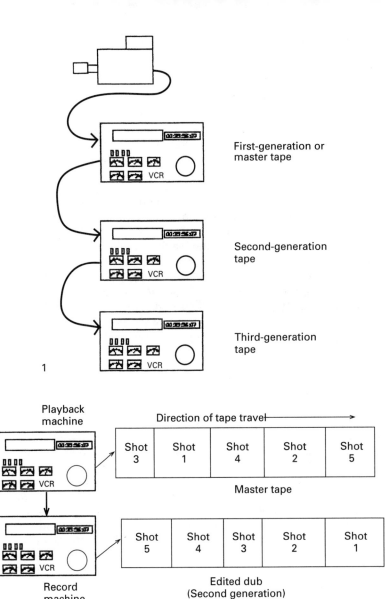

First-generation or master tape

Second-generation tape

Third-generation tape

1

Playback machine

Direction of tape travel ⟶

| Shot 3 | Shot 1 | Shot 4 | Shot 2 | Shot 5 |

Master tape

| Shot 5 | Shot 4 | Shot 3 | Shot 2 | Shot 1 |

Edited dub
(Second generation)

Record machine

2

1. **Tape generations.**

2. **Tape editing.**

The basis of editing is putting two shots together.

The Editing Process

It will be helpful to step through several edits made on a basic editing system just to become familiar with some concepts of the process. You begin with the master tape (containing the shots in nonsequential order), a blank tape, and the VTRs on which to run them: the playback VTR, which will run the master tape, and the record VTR, with which edits will be recorded on the blank tape. First, you find the beginning and end points of the first shot, also called edit points, and stop the master tape in the playback machine at the beginning of that shot. Preparing a tape for editing in this manner is called *cueing up* the tape. This is the first step in recording the first shot onto the blank tape.

The motors in each VTR must be at full speed and have achieved their full level of lockup before accurate recording can begin. Most modern editing machines take 3 to 5 seconds to reach full speed and lockup. Thus, rather than beginning the recording process at the first edit point, you must start the VTRs approximately 5 to 10 seconds before this point. This is called *prerolling* the VTRs. You stop the machines when the edit has been fully recorded onto the blank tape.

You then locate the edit points for the second shot on the master tape. The beginning of the second edit is cued up on the playback VTR, and the tape in the record VTR is stopped at the end of the first edit. Both machines are prerolled and then started in the playback mode. At the precise instant the playback and record VTRs reach, respectively, the beginning of the second edit and the end of the first, the record VTR is switched into the record mode. After the second edit is fully recorded, both machines are stopped and the entire process can be reviewed. You should be viewing shot one with a sharp, clean cut to shot two at the right moment.

You go through the same process for the rest of the shots in your show. This is how you build scenes into sequences, and sequences into final shows.

The process just described concerns by far the most basic edit possible and uses the most rudimentary equipment. The more advanced and precise the editing systems and methods are, the more complex the means by which video information is organized, accessed, and transferred into a sequence.

In a very basic system such as that described, the operator may unscientifically gauge when to make cuts and punch the specific button where he or she wants material to be copied. In advanced systems, such tasks are electronically programmed into a computer and precisely executed according to extremely accurate timing systems.

On longer videos especially, the editing process can be quite complicated. A crucial part of most editing tasks is the editing script or EDL (edit decision list). An EDL is a list of the shots you want to use, in the order in which you want to use them. The EDL will include the beginning and ending points of each shot. In very basic systems, these points may be specific visual or aural references and you may need to manually time the

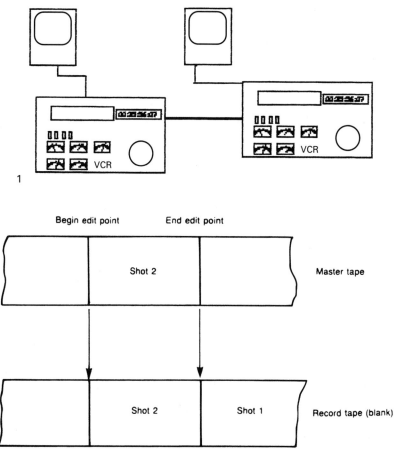

Begin edit point   End edit point

Shot 2   Master tape

Shot 2   Shot 1   Record tape (blank)

1. **Manual editing.**

2. **A typical video edit. A segment of tape is recorded from a master tape to its proper position on the record tape.**

length of each edit. Advanced systems have special coding systems by which you can designate the edit points.

Although creating an edit script may require a good deal of time, having a good script before you begin to make the edits will save you a great deal of effort. Ideally, most of your editing decisions will have been made ahead of time by reviewing the tapes, so that you will have less to worry about when you sit down to actually make the edits.

Types of Edits

Assemble edits

There are two basic types of edits: *assemble edits* and *insert edits*. With assemble edits, you start with a blank tape on which you will record. When you make the edits onto that tape, everything is recorded from the master tape: audio, video, and control track. Since the control track is coming off the playback machine rather than from the sync generator, the interval between pulses may be off (because of time base error), and the pulses themselves will be degraded (because of the generation drop). Because of these problems, the images may break up on the monitor when the edit is played back. The advantage of assemble editing is that it takes no advance preparation of the tape, but you may experience problems in getting good, clean edits.

Insert edits

Insert edits, on the other hand, require some preparation of the recording tape before editing begins. To do insert edits, a control track must be laid down first. This is usually done by placing a blank tape into a VTR, punching up black on the switcher, and putting the machine in record for the entire length of the tape. This will record a black video on the video tracks, and lay down the control track pulses from the sync generator.

Now you are ready to do insert edits, which give you a couple of advantages. First, you can edit using the procedure outlined above, but since the machine is using good, clean, consistent control track pulses, the edits are less likely to break up. Second, you can insert new material over old material. Assume that you had tape of a person talking about a terrific new invention. The audio information is interesting, but the video of a talking head is boring. So you take interesting video that you have of the invention (called *B-roll material*) and insert just that video over the boring talking head. Now, you have the interesting audio information as well as a visually interesting picture to look at. You could also insert new audio without changing the video you laid down previously. You can go either way. But the key thing to remember is that to do insert editing, the control track must be laid down on the tape you intend to edit to.

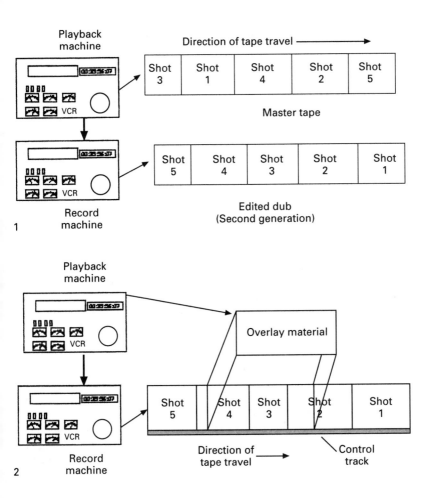

Playback machine

Direction of tape travel ⟶

Shot 3	Shot 1	Shot 4	Shot 2	Shot 5

Master tape

Record machine

Shot 5	Shot 4	Shot 3	Shot 2	Shot 1

Edited dub
(Second generation)

1

Playback machine

Overlay material

Record machine

Shot 5	Shot 4	Shot 3	Shot 2	Shot 1

Direction of tape travel ⟶

Control track

2

1. **Assemble edits**

2. **Insert edits.**

99

To be edited, video material must have a structure.

Editing Methods

Three basic methods are used to control and edit videotape: manual, control track, and SMPTE time code.

Manual editing

Manual editing was described earlier in "The Editing Process" section. You select and make edits based on visual or aural edit points. If, after prerolling both playback and record VTRs and placing them in playback mode, you decide the edit points are not coinciding as you'd like, you can abort the edit by not switching the record VTR to record mode. You can then change the amount of preroll on one or both tapes and try again. This is obviously a rather inaccurate and time-consuming way to edit, but sometimes it's the only choice.

Control track counters

A better method is to use a control track counter editing system. As noted earlier in "Video Recording Standards and Formats," vertical sync pulses are laid down regularly on the control track to stabilize the tape's play-back speed. A control track counter editing system uses these pulses as reference points to simulate and perform edits.

The control track counter is hooked up between the VTRs, although some systems have a counter built in. This editing process begins much as the manual method does. You select edit points on the playback and record machines. However, the editing system usually prerolls the machines for you and offers a preview option. If you select this mode, the system prerolls the machines and starts them, displaying first the output from the record machine (the previous shot) and then, at the correct edit point, switching to the playback machine's output (the current shot). This simulates the edit without actually making it. If the edit was good, you can have the system actually make the edit that was previewed.

This sounds like a pretty incredible machine, but it operates on a very simple principle. The heart of this editing system, the edit controller, merely counts control track pulses. So once the edit point is chosen, all it does is count pulses. When the controller backs up the machines for preroll, it backs both machines the same number of pulses, then rolls forward, counting the pulses, until it gets to the original edit points, at which point it makes the edit.

The controller ends the edit in the same way. Based on the end points that you predetermined, the controller measures the duration of the edit by counting the number of pulses. When you finally initiate the edit, the controller can recount the pulses and properly end the edit.

Since the controller is counting control track pulses, there had better be a control track laid down in the areas the controller engages. This is particularly important in assemble editing when you are adding a shot. The control track pulses at the very end of an edit (in the last few frames) may not be of good quality because the controller has already begun to end the edit. If your end point is exactly where you want the edit to end, the next edit's in-point may be either in an area of poor pulses or where there

100

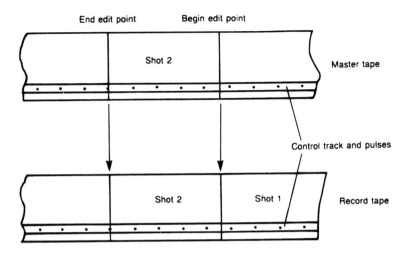

An edit made using a control track system. The system determines the length of the edit by counting control track pulses.

are no pulses at all. Therefore, when assemble editing, always let a scene continue longer than you need to; your next edit still can begin where you want it and you will have good pulses for the in-point.

Control track editing certainly is a vast improvement over manual editing, but there are still problems. The pulse counts may sometimes be unreliable as a means of measuring videotape material, since this isn't their original purpose. Control track editing is fine for some applications, but not accurate enough for many broadcast purposes.

In addition, because you cannot fully rely on control track to measure video material, you must keep a very careful editing script, recording audio or visual references for the edit points and manually timing each shot. Even if you carefully log such information, finding the shots you want to use can be very time consuming.

SMPTE Time Code Editing

One of the problems of the editing methods discussed earlier it that it is difficult to make an accurate editing script or edit decision list (EDL). An EDL is a list of the scenes you want to use in the order in which you want to use them. You would list the beginning point of each scene and the ending point of each scene. The problem is how to record, accurately, what the scenes are. If you describe the beginning and ending points you still have to search through the tape to find that particular scene. If you did several takes of the same scene, which one do you use? Or you could use the numbers on the tape machine counter. The problem here is that counters are not accurate. Every time you take a tape off the machine and put it back on the numbers will change. The solution to this problem is *SMPTE time code.* SMPTE time code uses one of the audio tracks or a special address track that many formats have to lay down a specific code number for each frame of video information on the tape. This code consists of a time: for example, 00:27:14:03 would be read as 0 hours, 27 minutes, 14 seconds, and 3 frames. This system works well partly because there are 30 video frames per second.

Since each tape would normally start at 00:00:00:00, the above address would be almost halfway into a 1-hour tape. The address is permanent. If the tape is taken off the machine and stored for a few years, 00:27:14:03 is going to be at exactly the same place on the tape coming out of the vault as it was when the tape went into the vault. Since most professional tapes are usually an hour long, the hours column of the time code is usually used to identify the tape. For example, a time code of 05:36:56:24 would normally be 36 minutes, 56 seconds, and 24 frames into tape number 5.

The first advantage of SMPTE time code is that it makes putting together an EDL much easier and faster. With SMPTE time code, when you find the beginning or ending edit point, you simply write down the time code number. You don't have to write down a description of the aural or visual cues that indicate the edit points. Thus, it's much faster to go through a tape and put together an editing script. This is particularly important if the video is very long and/or complex, such as a television program.

Editing by SMPTE time code is much more expensive, but it is also much more precise and reliable. For this reason, and because the code enables you to more easily organize and manipulate large amounts of material, most broadcast tapes are edited using time code.

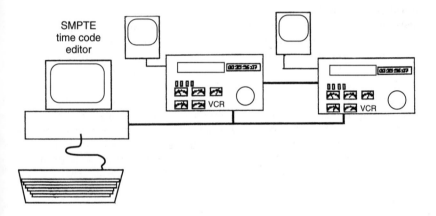

SMPTE time code editor.

Off-Line and On-Line Editing

SMPTE time code also makes on-line and off-line editing practical and economical. A top-of-the-line full-blown large-format editing suite (the system is so large it requires a room) can easily cost over $1 million to build, while an editing suite using lower quality small-format tape might cost less than a quarter of that. The rental fees of such suites will reflect the difference in costs. So it's going to cost a lot more (five or six times as much) to do all your work in the top quality suite than if you could do most of your work in the small-format suite and spend only a little time in the top quality suite. This is what on-line and off-line editing lets you do. To begin with, when you first record your show on large-format tape, you also record it on small-format tape and lay down the same time codes on the two tapes.

Off-line editing

During this phase you put your large-format tape in a storage vault and take your small-format tape to the less expensive editing suite and start developing your editing script by viewing the tape and selecting the order and edit points of the video segments you wish to join. As you develop the script, you also edit the tape. This will be your work print. Developing your editing script and work print is the most time-consuming part of the editing process. You don't really care about the quality of the tape since no one else is going to see it. Its primary purpose is to allow you to see how the edits fit together so that you can finalize your editing script. Putting together your editing script and your small-format work print that no one else is going to see is called *off-line editing.*

On-line editing

Putting together the large-format final tape is called *on-line editing.* After you finish the work print and the editing script, you retrieve your large tape from the vault and go to the expensive editing suite. The final, high-quality tape is easy to put together because during off-line editing you determined the specific end points and order of every edit. Since the time code numbers are the same for the small and large tapes, you can simply use your final editing script numbers to edit on-line.

Off line editing.

Video Compression

Digital video can take up an enormous amount of signal bandwidth, whether it is going down a cable or being transmitted over the air. Video compression allows that information to be compressed into a smaller bandwidth. When you are dealing with a limited amount of space (or bandwidth) the ability to compress more information into less space is a real advantage.

One of the great advantages of digital video is that it is much easier to manipulate and modify than analog information. Since the digital video signal is made up of a series of numbers, computer programs and mathematical formulas can be developed to add numbers, change numbers, or subtract numbers, thus adding video, changing video, or subtracting video. This idea is used in the process of video compression.

To understand how video compression works, it might be easier to deal with just two individual frames of video, one right after the other. In the first frame a race car is speeding around the race track. If we look at the next frame of video, which happens only $\frac{1}{30}$ of a second later, the two frames are almost identical. Things have moved a little, but in such a short period of time not very much can change. Using video compression, the system would send the first frame down the line where it would be both displayed and held in a memory. But instead of sending all of the information of the second frame, the system would send only the information that was different from that in the first frame.

On the receiving end, a computer program would take the information from the first frame that has not changed and combine it with the new information from the second frame. This new second frame would then be displayed and stored in the memory, where the process would start over again with just the new information for the third frame coming down the line. You can see that this process would greatly reduce the amount of information being transmitted at any one time, but it would require much more sophisticated equipment at both the transmitting and receiving ends.

A number of different mathematical formulas can be used in video compression. To make the system work, the transmitting end and the receiving end of the system must be using the same formula. These formulas are always a compromise. If you don't send enough information down the line, the final picture will not be as sharp and clear as it should be. On the other hand, if you are trying to save space, you don't want to send any more information than you have to.

Video compression can be used almost anywhere video information has to be stored (whether on tape or on some sort of computer memory device), or for transmitting information through cable or over the air.

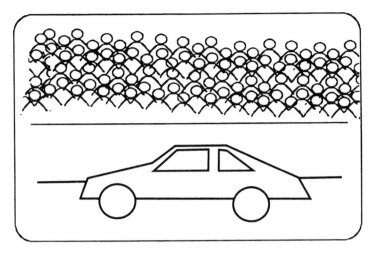

First Frame

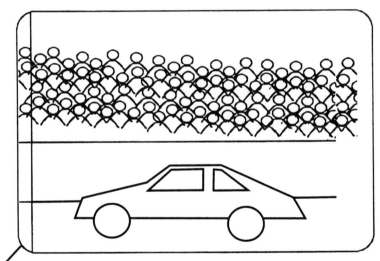

New information

Next Frame
1/30 second later

Video compression.

Computers help coordinate the editing process.

Editing by Computer

The real beauty of SMPTE time code editing is that special computers can do most of the work for you. The most sophisticated of these machines can control several tape machines and a programmable switcher at the same time. With more than one playback machine (provided they're properly time base corrected), you can dissolve, wipe, key, and do other effects between machines. All of this can be entered into the edit controller and it will do everything: Cue and preroll the machines, command the switcher to do special effects between the machines, and have the record machine do the actual edits. All you have to do is program the computer properly by entering the editing script containing all the scenes in sequence designated by SMPTE time code.

All of this sounds incredible, and it is, but this type of editing is fast dying out. In fact, videotape editing as we know it is dying. Control track editing is still commonly found in newsrooms around the country and will probably be with us for awhile. But traditional tape editing presents creative difficulties. This is especially the case for full-length programs.

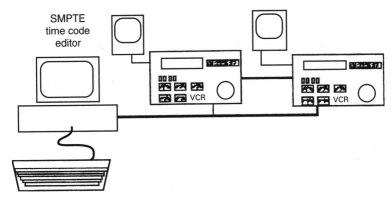

1

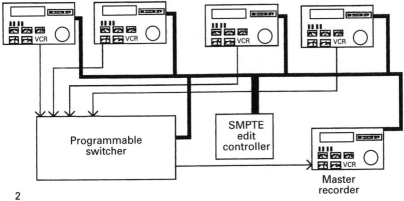

2

1. **SMPTE time code editor.**

2. **SMPTE time code editor running entire editing suite.**

109

Problems of Traditional Editing

There is a problem that limits flexibility and creativity in traditional editing. Once you have edited a sequence of scenes, you cannot go back and change one of those scenes without reediting the entire tape from that point on. Assume, for example, that you had edited a tape with four scenes in it. After looking at the tape several times you decide that the tape would be much better if the second scene started 2 seconds earlier. That will make the second scene and the tape 2 seconds longer. If you go back and reedit that scene and include the extra 2 seconds, the new scene will overlap and blank out the first 2 seconds of the third scene. As a result, you will have to reedit the third scene, which will now overlap the fourth scene. So you will have to reedit that. If you were to take that second scene and cut it down by 2 seconds, instead of adding 2 seconds, there would be unwanted material on the tape between the end of the new Scene 2 and the beginning of Scene 3. So, again, the tape would have to be reedited from the new end of Scene 2.

Imagine how much more complex and time-consuming the problem would be if you had to make a change early in a half-hour program that had 45 or more edits. This traditional process is called *linear editing* and, as you can imagine, is very time-consuming. If you are renting the editing suite, that additional time translates into an increased expense. Imagine how much time and money could be saved if you did not have to go back and completely reedit a tape after changing a scene.

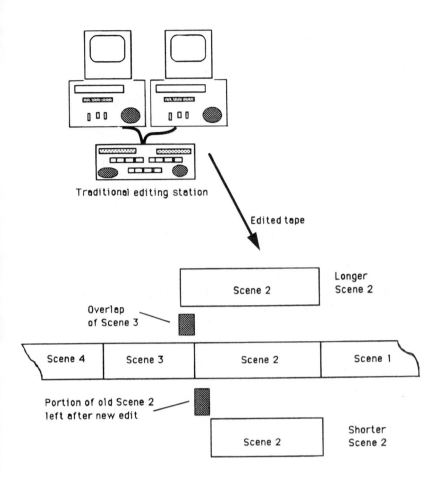

Traditional editing station

Edited tape

Longer
Scene 2

Scene 2

Overlap
of Scene 3

| Scene 4 | Scene 3 | Scene 2 | Scene 1 |

Portion of old Scene 2
left after new edit

Scene 2

Shorter
Scene 2

Problems of reediting a tape using traditional methods.

Nonlinear Editing

The advent of digital video and of new, more powerful small computers has brought about what is called *nonlinear editing.* This process allows the editor to go back and change or modify individual scenes at any time without the need to reedit the entire tape from that point on.

This process works because the raw video footage is digitized and stored in a digital memory. Using a powerful computer with specialized programming, individual shots and scenes can be called up from that memory. As the editor views the material, he or she can decide how long each shot should be and in what order the shots should be played back. When playing back an edited scene, the computer calls up the first shot chosen from its memory and displays the selected segment. It then calls up the next shot selected and displays it, and so on. What the viewer is actually seeing is selected segments of the computer memory displayed one after the other in a continuous manner. This will appear as a single uninterrupted sequence. In a sense the editor is not really editing video, but is creating a sophisticated playback program for the computer. The editor can put together the entire program in this way. If the editor wants to change a scene, he or she just goes back to that portion of the play-back and makes whatever additions or cuts are desired. When playing back the changed sequence, the computer will include the changes and play back the subsequent scene at the proper time.

In this way, the editor can do a "rough cut" of the entire show and then go back and fine-tune or polish the rough spots without having to reedit the entire segment. Once the editor is satisfied with the show, he or she can put together the edit decision list for the on-line editing of the show. Or you can just edit the final show. Early nonlinear editors did not produce good-quality pictures and could be used for off-line editing only. But today's nonlinear editors do outstanding on-line editing. In many cases, the off-line editing process is not used.

Since nonlinear editing first requires that the video be digitized and stored in a digital memory, there is an added step that requires some time that is not necessary for linear editing. This is why traditional editing may continue to be used in TV newsrooms for a number of years yet. Once the video is in memory, however, greater creativity and flexibility are possible. Whether there is a net saving of time using nonlinear editing will depend on the complexity of the project and how much reediting is required.

112

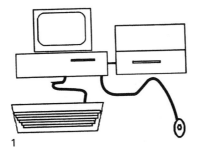

1

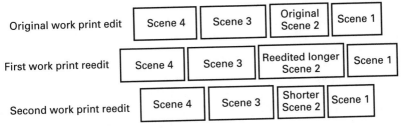

Original work print edit | Scene 4 | Scene 3 | Original Scene 2 | Scene 1

First work print reedit | Scene 4 | Scene 3 | Reedited longer Scene 2 | Scene 1

Second work print reedit | Scene 4 | Scene 3 | Shorter Scene 2 | Scene 1

2

1. Personal computer-based off-line nonlinear editing station.

2. Flexibility of nonlinear editing.

Computer Graphics for Video

Originating computer graphics

As you have already learned, digital video is really a series of numbers between 0 and 255 representing distinct voltages; the numbers are converted back to distinct voltages to form analog video. So, if you could sit down at a computer terminal and enter a series of about 250,000 numbers between 0 and 255, you would have just created a frame of video information. Needless to say, this would take a high degree of technical training and knowledge, as well as being a very time-consuming and tedious way to do things. What is needed is a method of creating digital video images that would be as simple as typing at a typewriter or drawing with a pencil and paper.

Interface between people and machines

A number of digital graphic systems have been developed that enable a person to take advantage of the opportunities offered by digital video. The capabilities of these systems range from producing simple letters and numbers to creating complex, detailed original illustrations and manipulating images on the screen. These systems provide an interface between the process of entering a series of numbers and the person who needs to get a job done.

Two things are necessary to make these interfaces successful. First, they must operate in a manner similar to what the user is already familiar with. If someone is introduced to a new piece of equipment that has several familiar features, he or she won't be as intimidated by it. But if a new piece of equipment is totally foreign, people will be more reluctant to try it out. Second, this new equipment needs to be flexible. The more that can be done with it, the more attractive it will be.

1. To effectively create a picture (or frame of video) on a computer, you might have had to enter thousands and thousands of values.

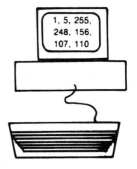

1

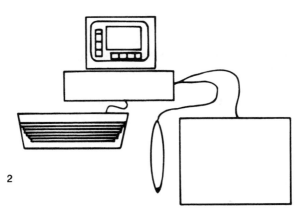

2

2. Fortunately, computers can translate our general commands into these thousands of specific values. We deliver these commands to the computer in simple ways, such as through a keyboard, an electronic "pencil and paper," or even voice.

Character Generators

Character generators (CG) were the first digital graphics units. A CG looks something like a computer terminal or a funny-looking typewriter. It is, in fact, intended to be used like a typewriter. Many of the editing features now used on electronic word processing systems were first used on CGs. CGs allow you to type titles onto the TV screen. If you want to make credits for who wrote, produced, directed, and starred in the show, use the CG. If you're interviewing someone and you want to flash his name up on the screen so the viewers know who he is, use the CG.

The earliest CGs were very simple. They only produced white lettering on black background for use as keys. There were one or two type sizes, and only one or two type styles (fonts). But now you can do white on black lettering, or you can give the letters any color you want and produce a colored background. You can make the letters an outline, or solid, or solid with an outline. A wide variety of sizes and fonts are available; the range is amazing.

To be most effective, character generators also need to have an extensive memory. CG memories are measured in pages. One complete video frame of information is a page. If you're preparing a show that has opening titles, names inserted during the show, and closing credits, you're going to need a lot of information. If the show is shot live, too much information will need to be typed as it is needed. However, you can enter all the information into the CG's memory before the show begins. Then it can be taken from memory when needed. Many CGs will have an internal memory of several pages and an external computer-like disk memory that will hold several hundred pages. Of course, what is really happening is that the CG is generating a series of numbers between 0 and 255 that are going to a D-to-A converter that emits an analog video signal from the CG. But the main control unit will still look very much like an electric typewriter's keyboard or a standard computer terminal.

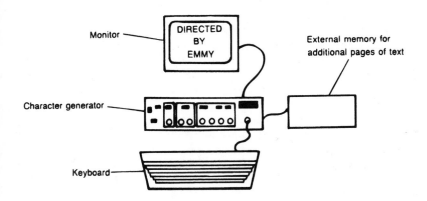

Monitor

DIRECTED
BY
EMMY

External memory for
additional pages of text

Character generator

Keyboard

Components of a typical character generator

Creating Imagery and Effects

Computer-generated imagery

As the name implies, computer-generated imagery (CGI) devices allow you to easily compose drawings or combinations of illustrations and text that can be manipulated and reproduced in a number of different forms, including on paper and as part of a video. These electronic graphic systems look very much like computers. There is the terminal, with the monitor (maybe two) and the disk drives, but there is also something that looks like a pencil and paper. It isn't really pencil and paper, of course, but an electronic palette and stylus. Most of the work is done with the palette and stylus so that an artist doesn't have to learn a lot about computers in order to use the system.

The palette looks like a flat rectangular piece of plastic with a wire coming out of it. The stylus looks like a rather fat ballpoint pen with a wire coming out of the top of it. The monitor displays a rectangular image area and menu boxes (lists of commands by which the computer receives instructions). When the point of the pen is touched to the palette, a cross usually appears on the monitor. As the point of the pen is moved across the palette, a corresponding line appears in the image area of the screen. So, while you draw on the palette, the image doesn't appear there, but rather appears on the monitor. You can choose different brush styles and textures, colors, and other details by simply touching the point of the pen to the appropriate menu box on the monitor. The keyboard terminal is used to enter text into the artwork, and the finished product can be stored and recalled from memory at will.

Each make and model of electronic graphics system will have its own features, but most will operate along the lines outlined here. Like the character generator, the system is storing an image as a long stream of numbers that will be converted to analog voltages at the output of the system.

Digital video effects

Digital video effects (DVE) systems enable a user to manipulate a digitized video signal and produce any of an incredible array of special effects, such as those described earlier. These devices can be incredibly complex, and the specific operation of different models can vary greatly. However, you should understand the basic concepts underlying their place in any video system. Remember that a DVE still serves as an interface between people and numbers. The video comes out of the switcher and into the DVE, where an A-to-D converter changes it to digital information. This digitized video is then modified by the manipulation of knobs, levers, and buttons on the DVE control panel. From there it goes to a D-to-A converter and back to the switcher.

Digital video has created vast new capabilities in TV production. Things that were unthought of 5 years ago are commonplace now. And any time the technology starts to become a little mind-boggling, it might help to remember that it's just a computer throwing around a bunch of numbers.

118

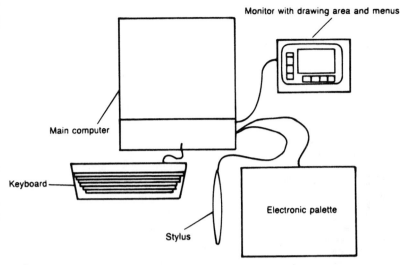

1

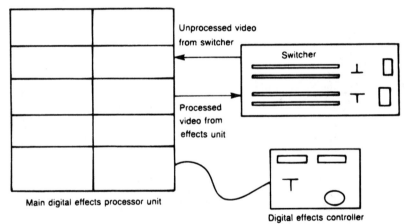

2

1. Typical components of a computer graphics system.

2. Typical digital effects setup.

Digital Videotape Recorders

There is great interest in digital videotape recorders because of their ability to go many, many generations without signal degradation or increased noise. The first two digital tape formats that found wide use in the industry are called D-1 and D-2.

D-1 component

The D-1 system is a component digital system. This machine won't even accept composite video. Only a digital signal can be fed into it. If you take the advantages of the analog component system discussed above and combine them with the advantages of the digital system discussed, you have amazing potential. This is exactly the type of system the D-1 digital videotape recorder was designed for. Since you have to run three channels of everything, this system is extremely expensive. It does, however, produce incredible pictures. The D-1 machine is particularly useful for very sophisticated production techniques, where one frame of video is laid down at a time on the tape, or where multilayered keys and special effects are needed. Professional video production houses, where top quality video is an absolute necessity, are the most common users of this machine. Not very many local stations make much use of the D-1 format.

D-2 composite

The D-2 composite system takes the single composite analog video signal and converts it to digital; it will also take a direct composite digital input. This digital signal is then laid down on the tape. The output of the D-2 machine is either composite analog or digital video. This allows the D-2 to be easily integrated with standard studio equipment. Several D-2 VTRs could also be hooked together for direct digital-to-digital editing. This is a much simpler and far less expensive system than D-1, yet D-2 is far superior to traditional analog video recording technologies. These machines will record in either the normal 3-by-4 aspect ratio or in a wide-screen format suitable to HDTV (see the discussion on that subject a little later).

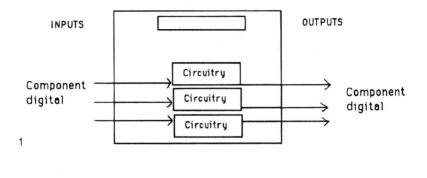

INPUTS   OUTPUTS

Component   Component
digital   digital

1

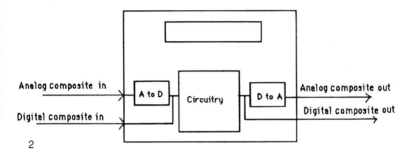

Analog composite in   A to D   Circuitry   D to A   Analog composite out

Digital composite in   Digital composite out

2

1. **D-1 component digital VTR.**

2. **D-2 digital tape.**

There is a great variety of digital tape formats now available.

Newer Digital Tape Formats

The D-1 component tape format has become the standard for top quality video production. It is the standard to which all others are compared. Since the introduction of D-1, a wide array of other digital videotape formats have been introduced. They range from using 6-mm-wide to 19-mm-wide tape; some use 3× sampling and some use 4× sampling; some use 4:1:1, while others use 4:2:2 encoding; and so on. Most are switchable between a normal aspect screen (4:3) or wide screen (16:9). Don't make the mistake of thinking that if something is in wide screen that it is high definition, because it may not be. In fact, if it is switchable between normal and wide screen, it is probably standard definition digital (SDTV). Almost all of these formats will have at least four audio tracks, some will have more. They will all have a separate address track for time code. What you choose to use will be determined by how much flexibility you need, what kinds of effects you will be doing, what level of picture quality you need, and how much you can afford to spend.

You will probably be better off picking a lower quality format and standardizing on that throughout your facility than trying to mix several formats in one system.

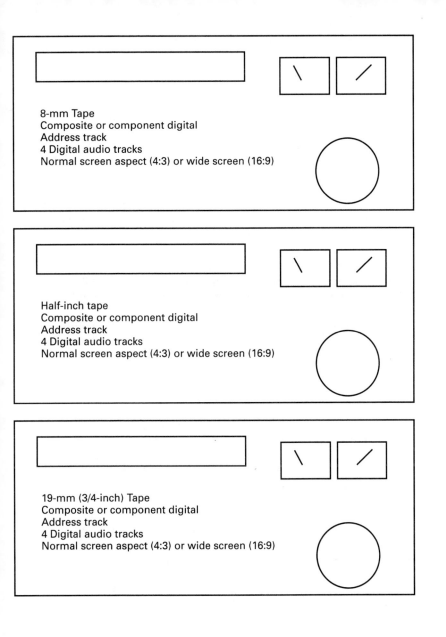

8-mm Tape
Composite or component digital
Address track
4 Digital audio tracks
Normal screen aspect (4:3) or wide screen (16:9)

Half-inch tape
Composite or component digital
Address track
4 Digital audio tracks
Normal screen aspect (4:3) or wide screen (16:9)

19-mm (3/4-inch) Tape
Composite or component digital
Address track
4 Digital audio tracks
Normal screen aspect (4:3) or wide screen (16:9)

Various digital videotape formats.

Digital Video Servers

Problems of videotape

Videotape recorders, in one form or another, have been around for 45 years and have served the industry well. They are used in all aspects of video production: acquiring original footage, editing, and playing back programs and commercials. The biggest problem with videotape machines is that they break down frequently and require a lot of maintenance. If a tape machine breaks down instead of showing a scheduled commercial, the station won't get paid for that commercial. Broadcasters would welcome a more reliable machine with equal or better picture and sound quality, if it doesn't cost too much.

Video servers

That machine is here and it is the *digital video server.* A video server very much resembles a group of computer hard disk drives. With digital compression it is possible to squeeze a great deal of video and audio information onto hard drives. One big difference between computer hard drives and video servers is that servers have several channels so that people can do different things at the same time. With your computer drive you can save a document or you can open something, but those tasks have to be done separately and only by your computer (unless they are networked, which is something else). With a server one person could be storing still images for later work, while at the same time someone else is playing back an edited piece of video from the same server. The number of people that can work simultaneously on a server is dependent on the number of channels it has. A two-channel server would allow two jobs to be done at the same time, whereas a four-channel server would allow four jobs to be done simultaneously.

Servers can be used for playing back commercials. They can be hooked up to a nonlinear editor and used for storage. They can be used for recording video in the studio. Just about anything you could do with a videotape machine can be done with a video server. Some studios are moving to a "tapeless" environment where videotape is not used at all. Everything is done on video servers. In most cases, videotape is still used in the field to acquire original footage. When back in the studio the tape will be dubbed to the server for editing and processing. However, a station might choose to use a digital videotape format. Some of these formats will allow you to transfer to a server at a faster speed than real time.

Servers are expensive, but the prices are coming down. This coupled with the fact that servers are much more reliable than videotape machines and that a station will not have to buy as much videotape makes digital video servers a real option for many stations.

124

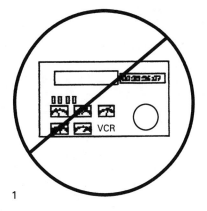

1

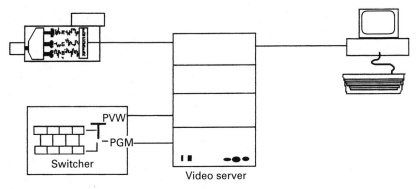

2

1. Tape is on its way out.

2. Digital video server.

Digital is pushing analog out of studios.

The Digital Studio

By this point you may be thinking that there is an awful lot of digital equipment out there, and you would be right. Digital cameras digitize the video information as soon as it comes off of the CCD and send it down the line. Digital distribution amplifiers can send that digital signal to several different destinations. Digital switchers allow you to select, process, and manipulate the digital signal. Character generators, computer-generated imagery equipment, nonlinear editors, and video servers are all digital pieces of equipment. If analog video isn't dead, it is certainly on its last legs. The adoption of a digital broadcast standard by the U.S. Federal Communications Commission (more on this in a few pages) ensures that more and more fully digital studios will be built in the next few years.

A completely digital studio is one in which you start with digital cameras and the signal remains in digital form right to and including the transmitter. The only piece of analog equipment that might remain in the studio or the home, for that matter, will be the CTR displaying the images. But if your final display will be on a plasma display screen, even that is digital. So the signal will be digital throughout the complete process unless it reaches a CRT where it will be converted to analog for final display.

The process of fully converting to digital will take place at different rates in different places. It is going to cost a lot of money; probably several million dollars for each station. Stations in the largest markets make the conversion more quickly and may rebuild their studios from the ground up. Stations in smaller markets will probably take more time and may purchase more and more equipment until it's a relatively simple matter to complete the conversion. It is harder to predict what will happen with schools and industrial production facilities. The cable industry, as of this writing, has not made a decision on digital. It is anybody's guess what will happen there.

Eventually, everyone will have to make the conversion to a completely digital system because in not too many years, analog video equipment will no longer be manufactured.

126

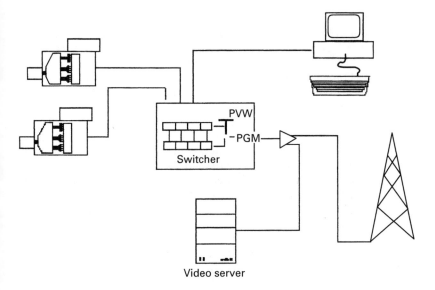

PVW

PGM

Switcher

Video server

The digital studio.

A whole new world of desktop video is opening up.

Open Architecture Equipment Versus Dedicated Equipment

Dedicated equipment is when each piece of equipment is dedicated to one specific job, such as switchers or character generators.

Many companies, instead of designing a piece of equipment to do one specific job, such as a character generator, are starting with a computer and writing special programs that will allow that computer to meet the special needs of video production. Special adapters have been built to allow the video signal and the computer to be integrated. This type of equipment is called *open architecture* equipment because the computer is not dedicated to any one specific use and is open to many uses, depending on the software (programs) running on the computer.

Using this approach, you can start with a high-level home computer and make that computer into a video graphics system, a character generator, a video switcher, a digital video effects unit, and an edit controller—all of this in one home computer. The quality of the work produced by some of these units is excellent. These systems are usually far less expensive than dedicated pieces of equipment.

The possibilities for open architecture equipment look wonderful. You could purchase one relatively inexpensive computer with the right software to do it all. A whole new world of desktop video, where individuals could do high-quality production out of their own homes, looks possible.

Open architecture equipment is definitely here to stay, but it may not live up to the predictions of some of the experts. Because you have one piece of equipment doing many jobs, you have to plan your tasks carefully in advance. Because the computer was not designed specifically for video production, the process of telling it what to do is more complicated. For example, to change cameras on a video switcher you simply have to press one or two buttons. With open architecture equipment, you would have to enter the instructions through the computer keyboard or with a computer mouse. As a result, dedicated equipment operates more quickly and the operator can change his or her mind and be more spontaneous with the equipment. Open architecture equipment may take over the editing suite, but dedicated equipment will have to be used for the foreseeable future for the production of live broadcasts, such as sports and news.

As for the prospect of desktop video, that is still quite a way off. Open architecture equipment is far from cheap. A decent open architecture system can cost $25,000 or more. You will need a quality camera and recorder to make and store the original images. While open architecture systems will allow more people to get into the video production business, it is not likely that many individuals will be able to afford that kind of expenditure without any expectation of significant financial return. The prospect of the average person being able to do quality video production at home is still a long way off.

128

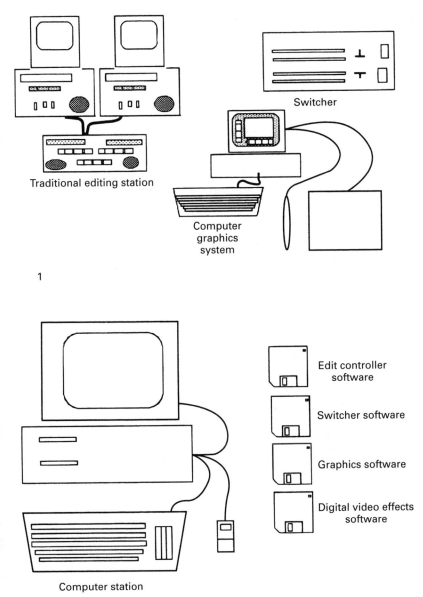

Switcher

Traditional editing station

Computer
graphics
system

1

Edit controller
software

Switcher software

Graphics software

Digital video effects
software

Computer station

2

1. **Dedicated video equipment.**

2. **Open architecture system.**

High-Definition TV

As good as many of the new systems are, they still leave something to be desired. You may have seen very-large-screen TVs and noticed that their pictures weren't as sharp and clear as smaller TV sets. That's because you've got the same 525 lines spread out over a much larger area. Any time you spread a given amount of information over a larger area, it's not going to look as sharp and clear, even in digital formats. This really demonstrates how much quality is lacking in a TV picture, compared to film. For many years there has been a great deal of research and development of TV systems that would produce a picture that approaches film quality. The result is called *high-definition television* (HDTV). The early work on HDTV used analog video, but more recent work has used digital video.

Production HDTV standards versus broadcast HDTV standards

There are two aspects of HDTV: production and broadcast. Production is the technology and equipment used to record and create programming. Broadcasting is the method of sending that programming through the air to your home. Under our old NTSC system the standards for both processes were the same. With the advent of digital technology, however, it is relatively easy to take material from one digital standard and convert it to another. You've discovered from reading this book that there is already a great deal of high-quality digital production equipment available: cameras, recorders, switchers, monitors, and so on. Much of this equipment is already commonly used in the production of video programming.

An alliance of researchers and manufacturers has worked to develop an HDTV broadcast standard for the United States. In April 1998 a new digital broadcast standard for the United States was announced by the Federal Communications Commission. This standard will be discussed in more detail in the next few pages, but we can look at some aspects of it now. The new standard is really a protocol that will allow several systems to be used at the same time. This makes possible the use of both standard definition digital television (covered in the next section) and high-definition television.

The high-definition system will be a digital system that uses the same amount of bandwidth (6 MHz) as the current NTSC system. The new system will have a wide-screen format similar to motion pictures. The format's aspect ratio will be 9 units high by 16 units wide, compared to the NTSC's 3 units high by 4 units wide. Some formats will use 1,080 interlaced scanning lines with 1,920 pixels per line, whereas other formats will use 720 progressive scanning lines with 1,280 pixels per line. You will recall that the NTSC system uses interlace scanning, where the odd-numbered lines are scanned first and then the even-numbered lines are scanned, and finally the two fields are interlaced to make a frame. Some of the formats will continue to use interlace scanning while others will

130

16:9 Aspect ratio

1080 Scanning lines of 1920 pixels for each line
with 30 frames of interlace scanning
or 24 or 30 frames of progressive scanning.

or 720 Scanning lines of 1280 pixels for each line
with 24, 30, or 60 frames of progressive scaning

Formats for high-definition digital television.

use progressive scanning, where each line is scanned in order: 1, 2, 3, 4, 5, 6, and so on. There is a lot of discussion on this issue. Many feel that interlace scanning gives the best picture on large displays while progressive scanning makes it easier to integrate with computers. The illustration above summarizes the various HDTV formats.

Standard Definition
Digital Television

Broadcasters have had reservations about high-definition television because it will cost a lot of money to convert to HDTV and they had seen no apparent way to increase their revenues to recover that additional cost. With the development of high-quality digital compression it is now possible to digitize a standard NTSC video signal and compress it so that four or more such signals can fit into one standard NTSC 6-MHz bandwidth channel. This is what happens if you have one of the small mini-dish home satellite systems. They send you four channels of digital video that have been compressed into one 6-MHz bandwidth. Your receiver then converts it back to NTSC to display on your home TV. This is called *standard definition digital television* or SDTV. SDTV can be sent in either a wide-screen format (16:9) or in normal width (4:3).

Using this sort of system, broadcasters could send out four programs of sports and for movies they send you 8 to 10 movies using one TV channel. This would give them four shows, or data such as Internet services, or programming information and details to sell instead of one and they would have the chance to increase their revenues to cover the cost of converting to digital broadcasting.

When the Federal Communications Commission announced the new digital broadcast standard for the United States in April 1998, it allowed for both high-definition and standard definition digital television. No one can predict exactly what is going to happen in the future, but the best guess is that stations will use a combination of SDTV and HDTV in their broadcast day. Perhaps during the daytime hours they will provide four programs that they can sell to different groups of advertisers and during evening prime-time hours and for sporting events their programming will be in HDTV.

As a consumer, if you buy a new digital television receiver or set-top converter you will be able to receive and watch all of the digital formats.

132

4:3 Aspect ratio

480 Scanning lines

704 Pixels per line

30 Frames interlace scanning

60, 30, or 24 Frames progressive scanning

1

16:9 Aspect ratio

480 Scanning lines

704 Pixels per line

30 Frames interlace scanning

60, 30, or 24 Frames progressive scanning

2

1. **Normal aspect formats for SDTV.**

2. **Wide-screen formats for SDTV**

133

Eventually NTSC will go the way of the dinosaur.

The FCC's Plan for Digital TV in the Future

When the FCC announced the new digital broadcast standard in 1998, it also announced a plan for implementing the new standard. To broadcast in digital, stations will be given a second channel to use, usually in the channel 14 to 69 range. This will allow them to continue to broadcast in NTSC on their old channel and at the same time to broadcast in digital on their new channel. Stations in the largest 10 metropolitan areas had to start digital broadcasting very quickly. Some were broadcasting in digital by mid-1999. Other stations have more time, but all TV stations in the United States are supposed to be broadcasting in a digital format by the end of 2003. It is hoped that this will encourage consumers to purchase digital TV sets or set top converters. By the end of 2006 NTSC broadcasting is scheduled to end completely. At that time, broadcasters will give up their old NTSC channels and continue broadcasting in digital on only their new channel.

This schedule may well change. It is one thing for engineers to hand-build a digital transmitter for testing purposes and it is something else again for manufacturers to design and mass produce 1,600 U.S. transmitters in a short period of time. It simply may not be possible to get all of the television stations in the United States converted to digital transmission by 2003. And consumers may not be ready to buy new TV sets or converters by 2006. Although the schedule may change, the United States will eventually be on a digital broadcast standard and NTSC will be relegated to the museums.

Although this plan is for the United States, it is likely that the rest of the world will follow. In fact, Europe is somewhat ahead of the United States with their digital broadcast standard. Digital production equipment is already the standard worldwide. NTSC production equipment is getting harder and harder to find. Once manufacturers fill the need for digital transmission equipment in the United States they will seek to expand their markets elsewhere.

People working in the production side of television have had and are having to make the transition between NTSC and digital. The American consumers will also have to make that transition.

134

FOR THE YEAR 2003

All TV stations in the United States transmitting in digital on a second channel.
Digital SDTV sets and converters available to the public.
Digital HDTV sets available to the public.

FOR THE YEAR 2006

All TV stations in the United States stop transmitting in analog NTSC.
For consumers, only digital SDTV and HDTV sets are available.

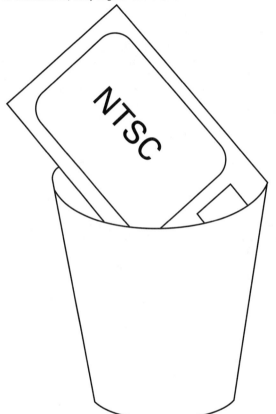

FCC plans for the future.

High-Quality Film-to-Video Transfer

The film and video industries are coming closer and closer together. Today it is common for motion pictures to be shot on film and have parts of it transferred to high-quality digital video. Once in digital form special effects can be done on computers. There is more than one method of converting from film to digital video, but we'll look at the device that is used most often.

Flying spot scanner

The most common of these high-quality devices uses a flying spot scanner. That simply means that, rather than using a movie projector, they use a CRT to illuminate the film. The figure at right shows the basic setup of a flying spot scanner. Notice that the only lens in the system is between the CRT and the film. The electron beam scanning the blank CRT provides the illumination, which is first picked up by the beam splitter and then divided into its component colors. This is an efficient way to transfer film to tape and gives excellent results.

There can still be a problem with the frame rates in that movies are shot at 24 frames per second and, as you know, most video is shot at 30 frames per second. But the flying spot scanner can be set to scan each frame alternately twice or three times. If you take 24 movie frames and scan half of them twice and the other half three times you will have 60 scans in a second. This will synchronize with the 60 video fields. In addition, some film directors who know they're going to use the unit will shoot their film at 30 frames per second in order to eliminate the conversion problem.

These film transfer machines do a wonderful job, but their high cost and the fact that local stations don't use film means that you won't find flying spot scanners at local stations. Large production houses in major metropolitan areas may have them as might companies that specialize in transferring film to video.

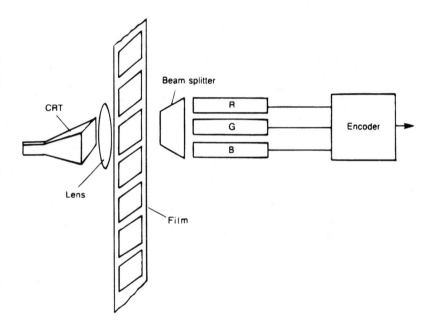

CRT

Lens

Beam splitter

R

G

B

Encoder

Film

Flying spot scanner.

137

Patch Panels

Patch panels are extremely basic items of equipment that are not always easy to understand. Patch panels help interconnect the video system components. For this reason, to use them successfully you must understand all of the other components and be able to carefully reason and evaluate possible connections and their results.

Because patch panels are such a vital piece of equipment, this special section both explains their use and offers some basic exercises. Remember that, although you will grasp the fundamentals by reading this text, to become fully comfortable with patch panels you should practice on your own with real equipment.

What patch panels do

Patch panels are routing devices that allow you to take the signal from one place to another. They don't change, convert, or do anything else to the signal. The panel is made up of a group of connectors (jacks), and if you want to hook up two pieces of equipment, you simply plug a cable into the output of one piece of equipment and into the input of another.

Patch panel components

Here's a quick example. You will be dealing with three pieces of equipment: (1) the panel itself, with holes (jacks) in it (each jack will be labeled); (2) the patch cable, which is made up of a video cable with a connector (plug) at each end, designed to fit into the jacks; and (3) a hairpin, little more than a U-shaped patch cable encased in a small block of plastic. It is used to connect jacks adjacent to each other on the patch panel. The first figure is a very small patch panel. It has jacks for only six inputs and outputs. Note that this particular patch panel has all outputs along the top and all inputs along the bottom, but not all patch panels are set up this way.

Patching example

Assume that you want to record the output of camera 2 on both VTR 1 and VTR 2. You would hairpin camera 2 into DA 1; then you would take the first DA output and hook it into the input of VTR 1 and the second output of DA 1 and hook it into the input of VTR 2. The panel would appear something like the last figure.

138

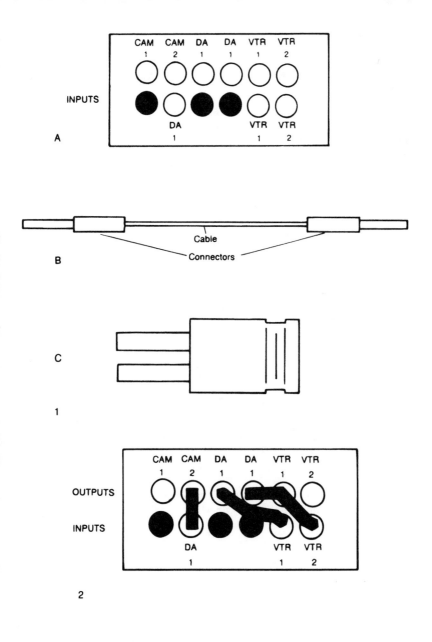

1. Patch panel components. (A) Simple patch panel; (B) patch cable; (C) hairpin.

2. Diagram of the patching example.

Patching Rules and Procedures

Patching rules

A couple of basic rules should be kept in mind when using patch panels:

1. Most pieces of equipment have only one output and one input. The DAs are the most notable exception to this rule. So if you need to take one signal to more than one place, you have to use a DA, as we did in the example above.

2. Outputs are always hooked up to inputs. Never output to output or input to input.

Patching procedures

There is also a procedure you can follow when trying to decide how to make a patch that may have up to five steps. The procedure itself is simple, but the questions it raises are not always that easy to answer:

1. What is your starting point? What piece of equipment is your signal coming from?

2. What is your destination? In other words, where do you want to go and where do you want the signal to end up?

3. Is there anything in between? Does any equipment need to go between the starting point and destination pieces of equipment? If the answer is no, then just patch between the pieces of equipment in steps 1 and 2 above. If the answer is yes, go to step 4.

4. What piece or pieces of equipment go between the starting point and the destination? If there is only one piece of equipment, you take the output from step 1 above to the input of the answer to this question. The output of that is taken to the input of the piece of equipment in step 2 above. If it takes more than one piece of equipment to answer this question, go on to step 5.

5. In what order? You need to decide in what order the intermediate pieces of equipment need to go, since the patch must be sequential.

140

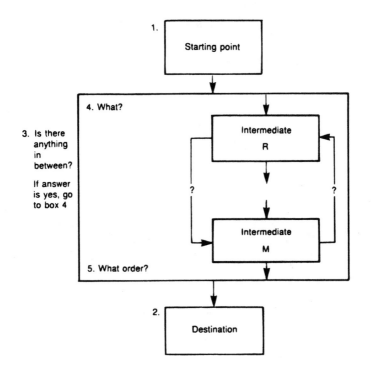

Steps for solving patch panel problems.

Explanation of a Small Patch Panel

Here's an example of a small patch panel; see if you can figure out some patches using the procedure on the previous page. As with the previous patch panel, this one has all the outputs along the top row and all the inputs along the bottom (first figure).

Go through all the jacks to make sure you understand the labels. Jacks 1 and 2 are the outputs of cameras 1 and 2. Output 3 is from the film chain. Output 4 is from the character generator. Output 5 is not being used. Outputs 6, 7, and 8 are from the program DA. In other words, they represent three identical outputs of the switcher. Output 9 comes from VTR 1, and 10 is the TBC output. Output 11 is not being used and 12 comes from VTR 2.

The lower jacks are the inputs. Jacks 13 and 14 are not being used. Jack 15 is the input to the CG. Jack 16 is the input to the production switcher that is labeled CG. Input 17 goes to the engineering switcher. (This is a switcher that allows the engineers to punch up whatever they want to look at in the system without interfering with the production.) Jack 18 is hooked up to a video connector in the studio in case a signal needs to be fed to some piece of equipment on the studio floor. Jack 19 is the input to the studio monitor so that the program can be seen in the studio. Input 20 goes to VTR 1 and input 21 goes into the TBC. Input 22 goes into the production switcher button labeled VTR 1, and 23 is the input to VTR 2. Jack 24 is not being used.

Normal setup

A number of these jacks should be hooked up in normal use (second figure). For example, you will probably want the CG to go to the switcher so that you can use titles during a production. So you would use a hairpin to connect jack 4 to jack 16. In addition, the program output (jack 8) is patched to jack 23 so that the program can be recorded on tape. You also want to be able to play back a tape through the switcher, but you need to time base correct the signal before you can do that, so jack 9 is patched to jack 21. To get the time base-corrected VTR signal into the switcher, you patch one end of the cable into jack 10 and the other end into jack 22.

Look at this patch panel for a minute and see if there is anything else you might want to hook up. It would be nice if the talent in the studio could see what was being recorded. What patch would you make in order to get that result? Go through the procedure again. What is your starting point and your destination? Since the program DA output is what is being recorded, that's what you want the talent to see, so that's your starting point. You already have a monitor in the studio, so that's your destination. Is there anything that goes between? The answer is no, so you can just patch the program output to the studio monitor. Since jack 8 is already being used, you can use either jack 6 or jack 7. Jack 7 has been hairpinned to jack 19 in the last figure. This would give you a setup of the patch panel that could be used in most normal shooting situations.

142

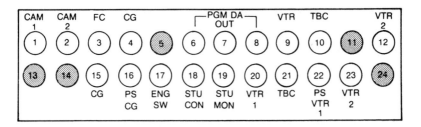

1

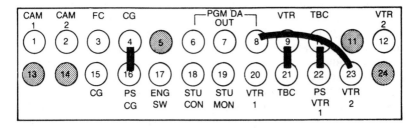

2

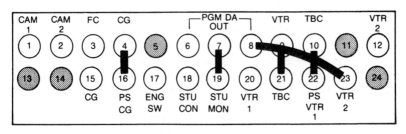

3

Patch panel example 1.

Simple Patching Exercises

Now, referring again to the basic setup shown in the preceding figure, assume that you have a different situation. The director wants to record the output of each camera on a separate VTR. That will allow him or her to select the best shots from each camera and edit them at a later time. This really presents two problems, since you have two starting points (cameras 1 and 2) and two destinations (VTRs 1 and 2).

It would be easier if you could pull all the cables and hairpins out of the patch panel and start with a clean panel. However, that is rarely done. Most patch panels are left in their normal configurations. Any new patches must be made from that configuration. Besides, if you pull all of the cables out of the patch panel, you might disconnect equipment that is being used elsewhere in the studio. Because of this, the problems presented here will assume that the patch panel is in its normal configuration.

In the problem presented above nothing needs to go between the cameras and VTRs, so you can hook them up directly. But first, you have to unhook the program output from VTR 2, since only one cable can be patched into a jack at a time (first figure).

Next you must solve a harder problem. Assume that your switcher has broken down. As a result, there's nothing coming out of any of the program DA outputs. But you want to edit between VTRs 1 and 2 and add titles in the process. Go through the procedure again. What's your starting point? How about VTR 1? Now what's your destination? VTR 2 will do just fine. Is there anything between? Yes, because you want to add titles. What? Since you need titles, the CG probably goes between. And it probably wouldn't hurt to stabilize the playback signal, so you probably want the TBC as well. In what order? Do you want to stabilize the video before it goes into or after it comes out of the CG? Unstable video certainly won't help the CG, so it's probably a good idea to stabilize it first. That leaves you with the flow diagram in the second figure.

Since the hairpin is already between the VTR 1 output and TBC input, you don't have to patch that. But you do need to pull out the hairpin between the CG output and the production switcher, and you also need to remove the cable between the program output and VTR 2 input. Then all you have to do is take the TBC output to the CG input, and the CG output to the input of VTR 2 (last figure).

Summary of patch panels

The patch panel that you've practiced on is intentionally small and simple so that you can develop a basic understanding of the equipment. Although patch panels in a typical TV station are much larger and more complex, the underlying principles are the same. Take your time when working on a patch panel and think the problem through carefully.

Patch panels tie the video system together. When you understand patch panels and are comfortable with them, you will truly understand the entire system.

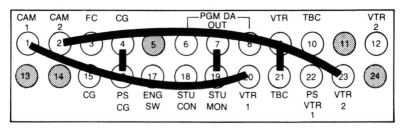

1

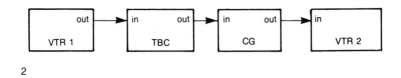

2

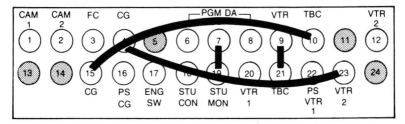

3

Patch panel example 2.

Further Reading

ALTEN, STANLEY:
Audio in Media, 5th ed. Belmont, CA: Wadsworth, 1986.

ANDERSON, GARY H.:
Video Editing and Post-production: A Professional Guide, 4th ed. Boston: Focal Press, 1998.

BROWNE, STEVEN E.:
Nonlinear Editing Basics: Electronic Film and Video Editing. Boston: Focal Press, 1998.
Video Editing: A Postproduction Primer, 3rd ed. Boston: Focal Press, 1997.

KALLENBERGER, RICHARD H., and KVJETNICANIN, GEORGE D.:
Film into Video: A Guide to Merging the Technologies. Boston: Focal Press, 1994.

MILLERSON, GERALD:
Effective TV Production, 3rd ed. Boston and London: Focal Press, 1994.
Lighting for Television and Film, 3rd ed. Boston and London: Focal Press, 1999.
Television Production, 13th ed. Boston and London: Focal Press, 1999.

OHANIAN, THOMAS A.:
Digital Nonlinear Editing: Editing Film and Video on the Desktop, 2nd ed. Boston: Focal Press, 1998.

ORINGEL, ROBERT S.:
Television Operations Handbook. Boston and London: Focal Press, 1984.

PAULSEN, KARL:
Video and Media Servers: Technology and Applications. Boston: Focal Press, 1998.

RUMSEY, FRANCIS, and WATKINSON, JOHN:
The Digital Interface Handbook, 2nd ed. Boston and London: Focal Press, 1995.

SCHNEIDER, ARTHUR:
Electronic Postproduction and Videotape Editing. Boston and London: Focal Press, 1989.

WATKINSON, JOHN:
The Art of Digital Audio. Boston and London: Focal Press, 1988.

WURTZEL, ALAN, and ROSENBAUM, JOHN:
Television Production, 4th ed. New York: McGraw-Hill, 1995.

ZETTL, HERBERT:
Television Production Handbook, 7th ed. Belmont, CA: Wadsworth, 1984.

Glossary

Additive colors   The color system that mixes colored light to create all of the various colors of the color spectrum.

Alternating current   An electrical circuit in which the flow of electrons reverses itself from negative to positive, from positive to negative, and back again at a regular rate.

Amperes (amps)   The unit of measurement for current.

Analog video signal   The varying voltages that make up the video information of a television signal.

Assemble edits   Edits that lay down all aspects of the signal: audio, video, and control track, all at the same time.

A-to-D converter   The circuitry that converts analog signal information into digital information.

Audio-follows-video switcher   A switcher that changes both audio and video sources with the push of one button.

Back porch   The portion of the waveform scan that represents the horizontal blanking just before the start of a new line of video.

Bandwidth   The amount of space available over the airwaves or through a cable for carrying information. A signal with more information requires more bandwidth to carry it.

Bit The smallest increment of computer memory represented by a 0 or a 1.

Black burst   A signal from the sync generator that includes all normal blanking and sync information along with black video.

Blanking   That time when the electron guns in the system are turned down to a low voltage so that they can return to the beginning of a new line or field.

Blanking pulses   Signals from the sync generator that signal the camera's electron gun to go into blanking.

Bus   A row of buttons on a switcher that allows a person to change between various video sources that are available in the system.

Byte   The smallest piece of computer memory that can be used as a distinct piece of information; made up of a group of bits.

Camera   A device that changes light images into a usable electronic signal.

Capstan servo   See **Vertical lock**.

Cathode-ray tube (CRT)   Television picture tube.

Character generator (CG)   A machine that creates words and titles for the TV screen.

Charge-coupled device (CCD)   A solid-state device used for changing light images into an electronic video signal.

Chroma   The color information in a TV signal.

Chroma key   A special effect in which a chosen color is replaced with video from another source.

Chroma key tracking   A digital effect that compresses the signal from a video source into the available chroma key window.

Color burst or **3.58**   The color reference inserted in every video line that determines how color information is to be interpreted.

Color subcarrier   See **Color burst**.

Color sync   See **Color burst**.

Complementary colors   The colors cyan, magenta, and yellow that are created by mixing parts of the primary colors (red, green, and blue).

Component switcher   A video switcher that deals with the individual color components (red, green, and blue) of the picture instead of the encoded composite video signal.

Component video   A video signal made up of the individual component parts as opposed to an encoded composite signal.

Composite video   The video signal made up of both the video and sync information.

Compressions   Digital effects in which the size and/or aspect of the picture is changed on the TV screen.

Control track   A track on the videotape used to help stabilize tape playback speed.

Control track counter editing controller   A device that controls videotape editing by counting the control track pulses on the tapes.

Current   The volume of electrons passing a given point at a given time, measured in amps.

D-1   A system of digital videotape recording that uses component video.

D-2   A system of digital videotape recording that uses composite video.

Dedicated equipment   Video equipment designed for a specific purpose, such as switchers, edit controllers, and character generators.

Deflection yoke   Electromagnetic coils around a CRT used to steer the CRT's electron beam as it sprays the picture across the screen.

Desktop video   The concept of private individuals being able to do low-cost, high-quality video production using personal computer-based production equipment.

Digital encoding ratios   A ratio that indicates the relationship between the amount of luminance (Y) and chrominance (B—Y and R—Y) information contained in a digital signal, for example, 4:2:2 or 4:1:1.

Digital video   A video signal that is made up of a series of assigned numbers rather than analog voltages.

Direct current   An electrical circuit where the flow of electrons moves in only one direction.

Distribution amplifier (DA)   A piece of equipment that produces multiple outputs identical to its input signal.

Drive pulses   Signals from the sync generator that control the scanning of the electron beams.

D-to-A converter   Circuitry that converts digital information into analog information.

Dubbing   The process of copying the electronic signal from one tape to another.

Dynamic tracking head   A videotape head that automatically aligns itself with the center of the video track on the tape for slow motion or freeze frames.

Electronic palette   A rectangular surface that represents the "paper" in a computer graphics system.

Encoding   The process of combining the chroma and luminance information into a single signal.

Fields   The complete set of odd- or even-numbered lines that, when interlaced, make up one video frame.

Flow diagrams   A diagram that uses geometric shapes and lines in place of equipment and wires to illustrate the interconnection of equipment.

Flying spot scanner   A film-to-video transfer device that uses the electron beam of a CRT to illuminate the film being transferred.

Fonts   The style of a particular typeface that is used on a character generator.

Frame   In the American system, two interlaced fields of 262.5 lines each that, when combined, make a complete picture of 525 lines. There are 30 frames per second in the NTSC (American) system.

Frame lock   A method of stabilizing videotape playback that tries to match an even field of the playback signal to an even field coming from the sync generator, and an odd field of the playback signal to an odd field coming from the sync generator.

Frame synchronizer   A device used to lock up nonsynchronous video signals to the main system.

Front porch   The portion of the waveform scan that represents the horizontal blanking at the end of a line of video.

Giga (G)   The abbreviation for billions (1,000,000,000). For example, 6 GHz would equal 6,000,000,000 Hz.

Gyroscopic time base error   Time base error that is created when a videotape recorder is moved perpendicular to the plane of the head drum's rotation.

HDTV   High-definition TV.

Head   The small electromagnetic device that lays down or picks up the information on a piece of recording tape.

Headwheel   The rotating disk on which the video heads are mounted

Helical A method of video recording that lays down video information at a slant to the tape's direction of travel; also known as *slant track recording*.

Hertz (Hz) A measurement of frequency equal to one cycle per second.

Horizontal blanking The period from when the electron guns are turned down to a low voltage at the end of a line until they are turned back up at the beginning of a new line.

Horizontal lock A method of stabilizing videotape playback that tries to match a horizontal sync pulse of the playback signal to each horizontal sync pulse coming from the sync generator.

Horizontal sync The signal from the sync generator that causes the electron gun to return to the other side of the screen for a new line.

Impedance A measurement of the properties that tell whether two or more circuits will interact well; measured in ohms.

Induction The process in which a circuit with a stronger magnetic field forces some of its signal into a circuit with a weaker magnetic field.

Insert edits Edits that use control tracks that have already been laid down on the tape.

Interlace scanning The process of taking a field of odd-numbered lines (1, 3, 5, 7, . . .) and combining it with a field of even-numbered lines (2, 4, 6, 8, . . .) to make a complete video frame (525).

Interpolation The mathematical process of creating new video information (pixels) from surrounding information.

Keys A special effect in which the signal from one video source "cuts" a hole into another video source.

Kilo (K) The abbreviation for thousand (5 K = 5,000).

Linear editing The traditional method of videotape editing in which one scene is laid down after another on tape. If any changes are needed after the edits have been made, the rest of the tape will have to be reedited.

Luminance The black-and-white portion of the video signal.

Luminance keys Keys in which the hole being cut is determined by the brightness of the video source.

Manual editing Editing that is completely done by a person without using an electronic editing controller.

Matte key A luminance key whereby the "hole" created by the key is filled with artificially created color from the switcher.

Mega (M) The abbreviation for million (3 M = 3,000,000).

micro (μ) The abbreviation for millionth (5 μ = 5/1,000,000).

milli (m)   The abbreviation for thousandth (200 m = 200/1,000).

Mismatch   Refers to impedance in which two pieces of equipment will not work well together.

Monitor   A TV set designed to display a straight video signal as opposed to a set designed to receive programs off the air.

nano (n)   The abbreviation for billionth (5 n = 5/1,000,000,000).

Noise   Unwanted electromagnetic static inherent in all electronic circuits.

Noncomposite video   Video information without sync information.

Nonlinear editing   A method in which video information is recorded into a digital memory where it can be called up a piece at a time in any order desired. Changes can be made in the edited program without reediting the entire show.

Nonsynchronous   A signal that is completely out of sync with the main system.

NTSC   The system of color television used in the United States and other parts of the world. The system uses 525 scanning lines and 60 fields with 30 frames per second (the field and frame rates have been rounded off). The name comes from the National Television Systems Committee, which was a group of industry experts that developed and proposed the system to the Federal Communications Commission in the early 1950s.

Off-line editing   The process of developing an editing script made up of SMPTE code numbers and a work print on small-format helical equipment.

Ohm (Ω)   The unit of measurement for both resistance and impedance.

On-air switcher   The switcher used to determine what goes to the transmitter. Usually an audio-follows-video switcher.

On-line editing   The process of editing the final, finished tape on large-format tape machines.

Open architecture   The concept of using computers with highly specialized programs to replace traditional dedicated equipment. Hence a computer could be a time base corrector, edit controller, video switcher, digital effects unit, and character generator, all in one.

Out of phase   When cameras show different colors during a transition because their color bursts are not matched.

Oxides   The coating on tape that allows signals to be recorded magnetically.

Pages   In a character generator or computer graphics system, one full screen of information that can be displayed at a time.

PAL   A color television system that was designed in Germany to overcome some of the problems of NTSC; it uses 625 scanning lines with 50 fields and 25 frames per second. This system is used in western Europe and many other parts of the world.

PAL-M   A color television system that is the same as PAL except that it is designed for countries that use a 60-Hz frequency for their AC power supply and therefore has 60 fields and 30 frames per second.

Patch panel A device that allows flexible routing of signals from one place to another.

Pedestal The black portions or areas of the TV picture.

Pickup tube A device that changes light images into an electrical signal.

Pixel Picture element.

Plasma display screen A video display device that uses electrically charged gases (plasma) to activate color pixels.

Positive interlace An interlace method in which the position of each line will be the same in every frame of video.

Primary colors In television, the colors red, green, and blue.

Proc amp Video processing amplifier, a piece of equipment that strips the distorted sync from a videotape playback signal and replaces it with clean sync. Most proc amps also allow the control of some of the video parameters, such as hue, video brightness, pedestal (black) levels, etc.

Quantizing The process of converting a sample of video information into a number.

Random interlace A system of interlace scanning in which the position of video lines scanned varies a little with each frame.

Resistance A measurement in ohms indicating the constraint to the flow of electricity.

Routing switcher A switcher used to route signals from one place to another.

Sampling The process of "grabbing" a piece of analog video information so that it can be quantized, or converted into numbers, for either processing or storage in memory.

Self-fill key A luminance key in which the hole cut is filled by the video that cut the hole.

Signal-to-noise ratio A comparison of the strength of the signal coming out of a piece of equipment to the internal noise created by that equipment.

SMPTE Society of Motion Picture and Television Engineers.

SMPTE time code A digital code laid down on tape that gives each frame of video a unique and unchanging address (03:38:52:04 would equal 3 hours, 38 minutes, 52 seconds, and 4 frames).

Software Computer programming that tells the computer how to process information.

Stylus An electronic device that acts as the pen, pencil, brush, etc., in a computer graphics system.

Switcher A device that allows a person to make a transition between video sources.

154

Sync   A signal from the sync generator that causes the electron guns to return to the beginning of a new video line or field.

Sync generator   A device that provides various sync signals (drive pulses, blanking pulses, and sync pulses) to keep all of the equipment in a video system working together.

Tape transport system   The mechanical system that pulls tape through a machine at an even speed.

Target   The photosensitive coating of a pickup tube.

Time base corrector (TBC)   A device for correcting time base error in videotape playback.

Time base error   The instability of a videotape playback signal created by the machine's inability to play back at exactly the same speed at which the tape was recorded.

Timing the system   Ensuring that all of the sync pulses from the various pieces of equipment arrive at the switcher at the same time.

Vectorscope   A piece of equipment that shows a graphic display of the color portion of the video signal.

Vertical blanking   The period when the electron beam is turned down at the end of a video field until it is turned back up at the start of a new field.

Vertical interval   The period when the electron beam is in vertical blanking.

Vertical interval switcher   A switcher that delays cuts between video sources until the entire system is in vertical blanking.

Vertical lock   A method of stabilizing videotape playback that tries to match the control track pulses of the playback signal to vertical sync pulses coming from the sync generator; also called *capstan servo*.

Vertical sync   The signal from the sync generator that tells the electron beams to return to the top of the screen for the start of a new video field.

Video compression   A technology that allows digital video information to be compressed into a smaller space, thereby requiring less bandwidth or memory space for transmission or storage.

Video servers   Computer-like hard drives specially designed to record and play back multiple channels of video information.

Videotape recorder (VTR)   A machine that records sound and pictures onto magnetic tape.

Volt   The measurement for the pressure of electricity.

Voltage   The pressure of electricity, measured in volts.

Watt   A measurement of electrical power.

Waveform monitor   A piece of equipment that shows a graphic display of the black-and-white portion of the video signal.

Window of correction The amount of time base error a TBC will correct, measured in video lines.

Y/C Y equals the luminance portion and C equals the chrominance portion of the video signal. A Y/C piece of equipment or system will keep the components separate as much as possible.

156